Springer Theses

Recognizing Outstanding Ph.D. Research

For further volumes:
http://www.springer.com/series/8790

Aims and Scope

The series "Springer Theses brings together a selection of the very best Ph.D. theses from around the world and across the physical sciences. Nominated and endorsed by two recognized specialists, each published volume has been selected for its scientific excellence and the high impact of its contents for the pertinent field of research. For greater accessibility to non-specialists, the published versions include an extended introduction, as well as a foreword by the student's supervisor explaining the special relevance of the work for the field. As a whole, the series will provide a valuable resource both for newcomers to the research fields described, and for other scientists seeking detailed background information on special questions. Finally, it provides an accredited documentation of the valuable contributions made by today's younger generation of scientists.

Theses are accepted into the series by invited nomination only and must fulfill all of the following criteria

- They must be written in good English.
- The topic should fall within the confines of Chemistry, Physics, Earth Sciences, Engineering and related interdisciplinary fields such as Materials, Nanoscience, Chemical Engineering, Complex Systems and Biophysics.
- The work reported in the thesis must represent a significant scientific advance.
- If the thesis includes previously published material, permission to reproduce this must be gained from the respective copyright holder.
- They must have been examined and passed during the 12 months prior to nomination.
- Each thesis should include a foreword by the supervisor outlining the significance of its content.
- The theses should have a clearly defined structure including an introduction accessible to scientists not expert in that particular field.

Onur Hamza Karabey

Electronic Beam Steering and Polarization Agile Planar Antennas in Liquid Crystal Technology

Doctoral Thesis accepted by
Technische Universität Darmstadt, Germany

Springer

Author
Dr. Onur Hamza Karabey
Institut für Mikrowellentechnik und Photonik (IMP), Fachgebiet Elektro- und Informationstechnik (ETiT)
Technische Universität Darmstadt
Darmstadt
Germany

Supervisor
Prof. Dr.-Ing. Rolf Jakoby
Institut für Mikrowellentechnik und Photonik (IMP), Fachgebiet Elektro- und Informationstechnik (ETiT)
Technische Universität Darmstadt
Darmstadt
Germany

ISSN 2190-5053 ISSN 2190-5061 (electronic)
ISBN 978-3-319-01423-4 ISBN 978-3-319-01424-1 (eBook)
DOI 10.1007/978-3-319-01424-1
Springer Cham Heidelberg New York Dordrecht London

Library of Congress Control Number: 2013944766

Printed on acid-free paper

Springer is part of Springer Science+Business Media (www.springer.com)

Parts of this thesis have been published in the following documents:

Journals

O. H. Karabey, S. Bildik, S. Bausch, S. Strunck, A. Gaebler, and R. Jakoby, "Continuously polarization agile antenna by using liquid crystal based tunable variable delay lines," *IEEE Transactions on Antennas and Propagation*, vol. 61, no. 1, jan., 2013.

O. H. Karabey, S. Gaebler, A.and Strunck, and R. Jakoby, "A 2-D electronically steered phased-array antenna with 2x2 elements in lc display technology," *IEEE Transactions on Microwave Theory and Techniques*, vol. 60, pp. 1297–1306, 2012.

O. H. Karabey, S. Bildik, S. Strunck, A. Gaebler, and R. Jakoby, "Continuously polarisation reconfigurable antenna element by using liquid crystal based tunable coupled line," *Electronics Letters*, vol. 48, pp. 141–143, 2012.

O. H. Karabey, M. Maasch, and R. Jakoby, "Stable satellite link by liquid crystal based phased array antennas," *ATZelektronik, Springer Vieweg*, vol. 04, pp. 46–50, 2012.

O. H. Karabey, M. Maasch, and R. Jakoby, "Stabile Satellitenverbindung durch Fluessigkristallbasierte Phasengesteuerte Gruppenantennen," *ATZelektronik, Springer Vieweg*, vol. 04, pp. 300-306, 2012.

International Conferences and Workshops

O. H. Karabey, S. Bausch, S. Bildik, S. Strunck, A. Gaebler, and R. Jakoby, "Design and application of a liquid crystal varactor based tunable coupled line for polarization agile antennas," in *Microwave Conference (EuMC), 2012 European*, 2012.

O. H. Karabey, F. Goelden, S. Gaebler, A.and Strunck, and R. Jakoby, "Tunable loaded line phase shifters for microwave applications," in *Microwave Symposium Digest (MTT), 2011 IEEE MTT-S International*, 2011.

O. H. Karabey, B. Saavedra, C. Fritzsch, S. Strunck, A. Gaebler, and R. Jakoby, "Methods for improving the tuning efficiency of liquid crystal based tunable phase shifters," in *Microwave Integrated Circuits Conference (EuMIC), 2011 European*, 2011.

O. H. Karabey, S. Bildik, C. Fritzsch, S. Strunck, A. Gaebler, R. Jakoby, and A. Manabe, "Liquid crystal based reconfigurable antenna arrays," in *32nd ESA Antenna Workshop on Antennas for Space Applications*, 2010.

O. H. Karabey, F. Goelden, A. Gaebler, and R. Jakoby, "Precise broadband microwave material characterization of liquids," in *Microwave Conference (EuMC), 2010 European*, 2010.

O. H. Karabey, Y. Zheng, A. Gaebler, F. Goelden, and R. Jakoby, "A synthesis technique for multiband tunable impedance matching networks with optimized domain," in *German Microwave Conference, 2009*, 2009.

Patent

O. H. Karabey, F. Goelden, A. Manabe, R. Jakoby: "Electronically steerable planar phased array antenna",
European Patent: EP 2 575 211 A1,
International Patent (PCT): WO2013045267A1.

For my estimable elder, E.A.,
and for his precious family.
His warm smile and kind words,
I will always remember.

Supervisor's Foreword

It is a great pleasure to introduce Dr. Onur Hamza Karabey's thesis work, accepted for publication within Springer Theses and awarded with a prize for an outstanding original work. Dr. Karabey joint my research group for Microwave Engineering after finishing the "Information and Communication Engineering (ICE) International Master Program" at Technische Universität Darmstadt in September 2008. He started his doctoral study with a three-year scholarship of the DFG-Graduiertenkolleg 1037 "Tunable Integrated Components in Microwave Technology and Optics" and completed it with an oral defense on 5th April 2013.

Since about a decade, researchers around the world are searching for appropriate enabling technologies to realize devices and components for advanced wireless systems at radio and microwave frequencies, having smart hardware functionalities, either for frequency tuning and adaptive impedance/power matching or electronic polarization-tuning and beam-steering. One promising approach, as presented in this thesis, is the microwave Liquid Crystal (LC) technology developed by my group at TU Darmstadt together with the liquid crystal division of the Merck Company in Darmstadt, where high-performance LC materials have been specifically synthesized for microwaves. This microwave LC materials feature low dielectric loss above 10 GHz as well as continuous and quasi-powerless tunability in terms of relative permittivity. By adapting the well-established LC display technology beyond optics, Dr. Karabey investigated in the frame of his thesis, different novel concepts for low-cost, low-profile (flat and compact) electronic beam-steering, and polarization-agile antennas. His comprehensive study focusses on 360° variable delay-line phase shifters in LC technology, as key components. His systematic analysis includes theory, design, techniques, and the proof-of-concept of multi-layer structures in various planar waveguide topologies with controllable characteristics by using thin LC cavities as well as planar loaded-line approaches, using LC varactors as tuning elements. For the very first time, Dr. Karabey successfully demonstrate the full capability of these novel approaches and the LC technology by integrating LC-based delay-line phase shifters in spiral geometry beneath each antenna element with physical size of about 0.55 $\lambda_0 \times 0.55\ \lambda_0$ (λ_0 is the wavelength in air) of a full 2D-beam-steering phased array and two new polarization-agile antennas. All these reconfigurable antenna demonstrators operate at Ku-/Ka-band and consist of 2 $\times$ 2 microstrip patch antenna elements with an overall thickness of 1.5 mm, only.

Dr. Karabey's thesis includes significant original scientific contributions representing considerable advancement in this field of electronically reconfigurable antennas based on LC technology, which he published as first author in top journals and international well-recognized conferences. Moreover, with his innovative ideas proposed in this thesis, Dr. Karabey is the main inventor of two patent applications and won the "Ideenwettbewerb 2011" of Technische Universität Darmstadt.

Utilizing automated manufacturing techniques analogy to LC display technology, Dr. Karabey's scientific work enables to fabricate low profile, low cost, high-gain beam-steering, and polarization-agile antennas, able to adjust dynamically its beam and polarization by electronics. This will have some potential for future portable and mobile applications, e.g., for airplanes connecting satellites or for terminals integrated into the body (cover, rooftop) of laptops, automobiles, ships, or boats, receiving broadcast or onboard Internet access if the antenna can electronically connected to a satellite constantly.

Darmstadt, 4 June 2013 Rolf Jakoby

Acknowledgments

This dissertation presents my research performed as Ph.D. candidate in the Institute for Microwave Engineering and Optics at Technische Universität Darmstadt. I would like to start the dissertation by acknowledging to people, who contributed to my research intangibly.

In the first place, I would like to thank my supervisor Prof. Dr.-Ing. Rolf Jakoby. Doubtless, this work could not be performed without his valuable guidance and trust. He always encouraged to take responsibilities and provided a flexible working environment, which also developed my personal skills. In the second place, I would like to thank Prof. Dr. Ozlem Aydin Civi. During my undergraduate studies, she taught me in antennas and now, she takes part in as a co-examiner in my Ph.D., which is dedicated for tunable antennas.

I am indebted to my former colleagues Dr. Stefan Mueller and especially to Dr. Felix Goelden for inspiring me for the topics studied in this work. These topics were, of course, advanced with fruitful discussion with my colleagues Alexander Gaebler, Sebastian Strunck, Saygin Bildik, Matthias Maasch, Christian Mandel, Carsten Fritzsch, Matthias Hansli, Dr. Yuliang Zheng, Dr. Christian Damm, and Dr. Holger Maune. My workload is reduced considerably with helps of Wenjuan Hu and Ananto Eka Prasetiadi with productive teamworking. This work became a reality with practical supports of Peter Kiesslich and Andreas Semrad in the workshop and Karin Boye in the clean room. I am also thankful to Arshad Mehmood for reviewing my dissertation.

Furthermore, my special thanks go to

- Atsutaka Manabe and his colleagues in Merck KGaA, Darmstadt because of the trustworthy cooperation and providing liquid crystal samples,
- Prof. Dr. Philippe Ferrari and Assoc. Prof. Dr. Anne-Laure Franc from the IMEPLAHC Microelectronics, Electromagnetism and Photonic Institute, Grenoble, with whom the liquid crystal technology is innovatively applied for slow wave phase shifters at W-band frequencies,
- Prof. Dr. Wolfgang Haase and Dr. Artsiom Lapanik from the Eduard-Zintl-Institut fuer Anorganische and Physikalische Chemie at Technische Universität Darmstadt for helping me to get insight into liquid crystal chemistry,
- Dr. Serhend Arvas from the Sonnet Software Inc., North Syracus for teaching me to use simulation tools efficiently, regardless of the distance.

Last but not least, I would like to thank my relatives, parents, and parents in love for supporting and motivating me in uneasy and long Ph.D. journey. Special thanks go to my darling wife and four month old son. They supported me morally and shared my difficulties in any case.

Neu-Isenburg, 23 January 2013 Onur Hamza Karabey

Contents

Symbols and Abbreviations

α_p	Electric polarisability
χ_e	Electric susceptibility
χ_m	Magnetic susceptibility
$\Delta\phi_\beta$	Differential phase shift
ℓ_{physical}	Physical length of a device
η_τ	Tuning efficiency of a liquid crystal device
η_{LC}	Material tunability
η_{VDL}	Variable delay-line efficiency
Γ	Reflection coefficients
γ_{LC}	Rotational viscosity of the liquid crystal molecules
k	Magnetic polarizability
μ_0	Permeability of free space, $4\pi \times 10^{-7}$ N/A^2
ω	Angular frequency
$\overleftrightarrow{()}$	Tensorial matrix of a given property in ()
$\parallel$	Parallel alignment with respect to an electromagnetic field
$\perp$	Perpendicular alignment with respect to an electromagnetic field
ϕ_ℓ	Electrical length
ϕ_β	Phase shift
$\phi_{\ell,\text{NTL}}$	Electrical length of a not tunable line
σ	Conductivity
$\tan\delta$	Material (dielectric) loss tangent
τ_{LC}	Material efficiency
IL	Insertion loss
N	Molecular number density
Θ_{m}	Main beam direction
$\underline{\gamma}$	Complex propagation constant
$\underline{\mu}_r$	Complex relative permeability
$\underline{\varepsilon}_r$	Complex relative permittivity
$\underline{\varepsilon}_{\text{eff}}$	Complex effective relative permittivity
S_{ij}	Scattering parameters

ε_0	Permittivity of free space, 8.85 x 10^{-12} F/m
ε_r	Realitive permittivity
$\vec{n}$	Director
$\vec{D}$	Dielectric displacement
$\vec{E}$	Electrical field
$\vec{H}$	Magnetic field
$\vec{P}$	Macroscopic polarization
$\vec{p}$	Induced dipole moment
c_0	Speed of light in vacuum $\approx$ 3 x 10^8 m/s
f_0	Operating frequency
h_{LC}	Height of the liquid crystal layer
K_{ii},	Elastic constant for the splay (i = 1), twist (i = 2) and bend (i = 3) deformations
S	Order parameter
t_{on}, t_{off}	Response times
T_{c}	Clearing temperature point
T_{m}	Melting temperature point
V_{Sat}	Saturation voltage
V_{th}	Threshold voltage
V_{b}	Bias voltage
Z_{Line}	Characteristic impedance of a transmission line
Z_{NTL}	Characteristic impedance of a not tunable line
Z_B	Bloch impedance
Au	Gold metal
IMSL	Inverted microstrip line
LC	Liquid crystal
MUT	Material under test
Ni-Cr	Nickel-chromium alloy
NRW	Nicolson-Ross-Weir
NTL	Not tunable line
RF	Radio frequency
SLL	Side lobe level
UE	Unit element
UV	Ultraviolet light
VAR	Varactor
VDL	Variable delay line

Chapter 1
Introduction

1.1 Motivation

In modern wireless communication systems, there is a remarkable demand for high performance, multi-functional, cost-effective and compact devices. In this context, antennas are critical or even limiting components having impact on system performance, cost and compactness.

Regarding the system performance, *reconfigurability* of an antenna is a key issue since it enables adapting antenna's characteristic with changing environment and system requirements [Ber07]. For example, in mobile terminals such as automobiles, the reconfigurability refers to beam scanning or a beam forming antenna, which connects the mobile terminal reliably to a desired point, e.g. to a satellite. Hence, the antenna can provide on-board internet access and satellite television reception. In radar applications or in multiple input multiple output (MIMO) systems, the reconfigurability also includes polarization agility to improve a noisy connection [Ber07, AL92] , to enhance communication channel capacity [PD06] or to reduce polarization loss caused by atmospheric effects, for example rain drops, and by multipath propagation. Last but not least, the reconfigurability leads to frequency agility as well. For instance, tunable multiband antennas utilized in mobile phones feature independently tunable operating bands [Zhe11], which results in wider instantaneous bandwidths [Ber07].

Considering cost-effectiveness of a communication system, *emerging technologies* for tunable devices are of interest, which require low manufacturing costs together with high reliability and yield. Additional requirement is applicability of a chosen technology to higher microwave frequencies, especially for $f > 10$ GHz, if modern or even future communication systems are targeted. Technological approaches include radio frequency microelectromechanical systems (RF MEMS), semiconductor solutions and ferroelectrics such as barium-strontium-titanate (BST). Another approach is the usage of liquid crystal (LC) material beyond optics. LC is an anisotropic material resulting in tunability in terms of the relative permittivity. As its name suggests, the material can flow like liquid, which allows controlling the orientation of the molecules by using surface anchoring or external electric or mag-

O. H. Karabey, *Electronic Beam Steering and Polarization Agile Planar Antennas in Liquid Crystal Technology*, Springer Theses, DOI: 10.1007/978-3-319-01424-1_1,

netic bias fields. Indeed, this is necessary to exploit LCs as tunable materials. In contrast to BST material, the dielectric loss of the LC material, decreases with an increasing frequency, having high performances generally above 10 GHz [Mue07, BMS10].

The last requirement that is the compactness of a communication system is subject to antenna's low profile. In fact, the antenna profile depends on a chosen technology and also how innovatively an antenna concept is designed.

This work presents reconfigurable planar antennas realized in LC display technology, which advances the state-of-the-art further. The antennas feature reconfigurability in sense of electronic beam steering or polarization agility. Novel variable delay lines are studied systematically in multilayer planar, or printed, transmission line technologies, including coplanar waveguide, slot- and microstrip line technologies. Indeed, these transmission lines feature system-inherently much higher losses than hollow waveguide topologies, but because of their planarity, they are system-inherently low-profile and they can be realized rapidly by utilizing automated manufacturing techniques. Therefore, innovative antenna concepts are proposed, featuring a desired antenna reconfigurability as well as planarity and aesthetic appearance with a low profile. These antennas are realized by adapting the industrialized LC display technology for microwave devices [Goe09]. This leads to low manufacturing costs, if the antennas are even scaled in larger sizes.

1.2 Electronic Beam Steering Phased Arrays

Various services such as wireless internet, multimedia, communication and broadcasting services are provided by terrestrial point-to-multipoint systems and satellites, operating in the L-, C-, Ku-, Ka- or W-band. To provide these services also for mobile terminals, phased arrays have been studied intensively [PZ02, DJ93, Mou+08] for military and industry ground station applications, including mobile terminals such as airborne [TTO91], ship borne [Gei+11] or automobiles [Bag+10].

Most of the phased array antennas are controlled by using a hybrid tracking method, where the antenna is steered electronically in elevation plane and mechanically in azimuth plane [JKO00, Fak+09]. These architectures allow wide-angle scanning of flat arrays with a negligible gain loss [Tos08]. However, the antennas are heavy and occupy high volume because of the mechanical systems. For mobile terminals, not only the high performance but also the flatness and low-profile of the antenna are required. Similar to a car collision avoidance radar [Jon01], the phased arrays should be integrated into the structure of the mobile terminal, e.g. into the rooftop or body of a vehicle.

Planar phased array based on RF MEMS [Jun+06, Top+08, KPP08] and monolithic microwave integrated circuit (MMIC) [Hon+06, KJ08] have been demonstrated for Ku- and Ka-bands. RF MEMS systems offer a better performance compared to MMICs, because of lower insertion loss of the phase shifters [KP06]. Nonetheless, their practical implementation is challenged by the environmental conditions, and therefore, they require hermetic packaging. Hence, different microwave materials, including flexible liquid crystal polymer [KPP08] and low temperature co-fired ceramics (LTCC) [Gau+09], have been used to reduce the packaging cost and to improve the system integration of RF MEMS based antennas. Recently, the RF MEMS switches and a MMIC based low noise amplifier are combined in one block [Bag+09] and integrated into a Ku-band active phased array [Bag+10] for a cost-effective antenna production. Planar phased arrays can also be implemented by using tunable dielectrics. For instance, BST based phased arrays have been studied in [Rom+00, Saz+11b] . This technology offers relatively high performance for frequencies below Ku-band whereas their performances decrease considerably beyond this frequency because of increasing dielectric losses [Jak+04]. In [SHK09], an electronic beam steering phased array operating at Ka-band is presented. It includes a 4×4 tapered slot array and LC filled LTCC phase shifters. The main beam is steered $\pm 30°$ in one dimension only. This is due to the fact that the proposed phase shifter is relatively long and therefore, does not allow to feed each radiating element individually.

1.3 Polarization Agile Antennas

Modern communication systems demand for high information flow almost in all areas of communications. Achieving the requirements become more challenging when the fading rate is high, i.e. in urban areas. In this context, a polarization agile antenna is of interest, whose polarization state can be controlled from one polarization to another one, including linear and circular polarizations [GSZ06].

Polarization agile antennas can be used for many applications such as polarization diversity [Don+05], synthetic-aperture-radar (SAR) systems [AL92], RFID systems [Kos00]. In [AL92], for instance, it is shown that a polarimetric SAR provides more information about the target scattering mechanisms compared to those of a single-polarization SAR system. A technique known as the polarimetric whitening filter is used and horizontal-horizontal, horizontal-vertical and vertical-vertical polarization components are combined [Nov+97]. As a result quality of imageries is improved because more information is provided by using different polarization components. Hence, a dual polarization acquisition mode has been already installed to several current satellite SAR systems such as TerraSAR-X, COSMO-SkyMed [Kry+11]. For the near future, compact and cost-effective systems can be implemented by using a single polarization agile antenna instead of multiple antennas with different polarizations [GSZ06].

The approach for the polarization agility should be feasible and cost-effective for large antennas. The pronounced antenna functionality is mainly accomplished by alternating the current flow of either its feeding network [WR06, Li+10, Qin+10] or radiating elements [Sch+81, HD94, Hu+07, WR09, PC10]. For the former, tunable components such as phase shifters [ZYG04, DOW94] and coupled line [Fer+09] have been studied. These components can be included in the antenna layout in two possible ways. The first option is to place one tunable component for each radiating element as given in [PCD10]. The advantage of this topology is that the design of the feeding network of the array is relatively simple. However, for large arrays, the tunable component is supposed to be low-cost, compact and biased individually, which results in a complex biasing network. The second possibility is to use only one tunable component for the whole array like in [ZYG04]. This is a cost-effective solution and the biasing network design is simplified considerably. However, the design of the feeding network becomes challenging. For dual-fed radiating elements, for instance, two feeding networks have to be implemented for a *horizontal* (H) and a *vertical* (V) excitation of the radiating elements.

From the technological point of view mainly pin diodes [WR06, Li+10, Qin+10, Hu+07] and semiconductor varactors [HD94, Fer+09, WR09] (or even both approaches [PC10]) have been preferred as tunable components because of their high integration capability and performance. The proposed antennas in previous works can be efficiently used generally up to C-band because of technology limitations, i.e. low performance of varactors at higher frequencies. However, operating frequencies of modern communication systems shift to millimeter and even submillimeter wave frequencies because of the increasing demand for high data rate. Additionally, antennas implemented by using pin diodes can only be switched discretely from one polarization state to another one.

1.4 Structure of this Work

After the first indroduction chapter, Chapter 2, presents the fundamentals of LC materials in accordance with their usage in microwave applications. The material anisotropy is explained together with methods for controlling these behaviors. Additionally, basic knowledge on technology and working principles of LC microwave devices are presented.

The Chapter 3 describes three microwave material characterization methods and measurement results. It is aimed to determine the material parameters of both the LC materials and glass substrates. In the light of the results obtained in this chapter, antennas and their components are designed in the following chapters.

The Chapter 4 covers all essential aspects of a two-dimensional (2D) electronic beam steering phased array. As being a key component, LC based variable delay lines are focused. Several delay line topologies are realized and their measurement results are presented. Based on these studies, a phased array concept is proposed in LC technology for flat and low profile antennas. A prototype is fabricated as proof-of-concept and its measurement results are discussed.

The Chapter 5 presents two types of polarization agile antennas. Empirical studies are performed to utilize LC material tunability efficiently. Two antennas are realized as proof-of-concepts. The pronounced antenna agility is confirmed by the measurement results.

Finally, the Chapter 6 concludes significant achievements. Especially, the results obtained in this work are compared to the published results, to figure out the advancement of this work. An outlook is given as well for further investigations and enhancements of these results.

References

[AL92] Agard-Ls-182, *Fundamentals and Special Problems of Synthetic Aperture Radar (SAR)*. Technical report, NATO Research and Technology Organisation (1992), http://ftp.rta.nato.int/public//PubFullText/LS/AGARD-LS-182//AGARD-LS-182.pdf

[BMS10] S. Bulja, D. Mirshekar-Syahkal, Meander line millimetre-wave liquid crystal based phase shifter. Electron. Lett. **46**(11), 769–771 (2010)

[Bag+09] L. Baggen et al., Mems-4-mmic: Innovative rf-mems mmic components for agile rf-frontends. in *Proceedings of the International Conference Electromagnetics in Advanced Applications ICEAA '09* (2009), pp. 311–314

[Bag+10] L. Baggen et al., Compact phased arrays for mobile terminals. Proc. Int. Semicond. Conf. (CAS) **01**, 3–9 (2010)

[Ber07] J.T. Bernhard, Reconfigurable antennas. *Synthesis Lectures on Antennas* (Morgan and Claypool Publishers, Princeton, 2007)

[DJ93] A. Densmore, V. Jamnejad, K-and ka-band satellite-tracking, small reflector vehicular antenna system for mobile satellite service. in *Proceedings of the World European Conference Satellite Communications—ECSC-3*, pp. 117–121

[DOW94] S. Dauguet, E. Ongareau, J. Wyrwinski, Microstrip antenna with polarization switching. Microw. Opt. Technol. Lett. **7**, 36–40

[Don+05] Liang Dong et al., Simulation of mimo channel capacity with antenna polarization diversity. IEEE Trans. Wireless Commun. **4**(4), 1869–1873 (2005)

[Fak+09] M. Fakharzadeh et al., Fast beamforming for mobile satellite receiver phased arrays: theory and experiment. IEEE Trans. Antennas Propag. **57**(6), 1645–1654 (2009)

[Fer+09] F. Ferrero et al., A novel quad-polarization agile patch antenna. IEEE Trans. Antennas Propag. **57**(5), 1563–1567 (2009)

[GSZ06] S. Gao, A. Sambell, S.S. Zhong, Polarization-agile antennas. IEEE Antennas Propag. Mag. **48**(3), 28–37 (2006)

[Gau+09] W. Gautier et al., Rf-mems phased array antenna on low-loss ltcc substrate for ka-band data link. in *Proceedings of the European Microwave Conference EuMC* (2009), pp. 914–917

[Gei+11] M. Geissler et al. Phased array for maritime satcom applications. in *Proceedings of the German Microwave Conference (GeMIC)* (2011), pp. 1–4

[Goe09] F. Goelden, *Liquid Crystal Based Microwave Components with Fast Response Times: Material, Technology, Power Handling Capability*. Ph.D. thesis, Technische Universitaet, Darmstadt (2009)

[HD94] P.M. Haskins, J.S. Dahele, Varactor-diode loaded passive polarisation-agile patch antenna. Electron. Lett. **30**(13), 1074–1075 (1994)

[Hon+06] Seungpyo Hong et al., A multiband, compact, and full-duplex beam scanning antenna transceiver system operating from 10 to 35 ghz. IEEE Trans. Antennas Propag. **54**(2), 359–367 (2006)

[Hu+07] C.-H. Hu et al., Reconfigurable microstrip antenna with polarisation diversity and frequency agility. Electron. Lett. **43**(24), 1329–1330 (2007)

[JKO00] Soon-Ik Jeon, Young-Wan Kim, Oh Deog-Gil, A new active phased array antenna for mobile direct broadcasting satellite reception. IEEE Transactions on Broadcasting **46**(1), 34–40 (2000)

[Jak+04] R. Jakoby et al., Nonlinear dielectrics for tunable microwave components. in *Proceedings of the 5th International Conference on Microwaves, Radar and Wireless Communications MIKON-2004*, vol. 2, pp. 369–378

[Jon01] W.D. Jones, Keeping cars from crashing. IEEE Spectr. **38**(9), 40–45 (2001)

[Jun+06] C. won Jung et al., Reconfigurable scan-beam single-arm spiral antenna integrated with rf-mems switches. IEEE Trans. Antennas Propag. **54**(2), 455–463 (2006)

[KJ08] N. Khandelwal, R.W. Jackson, Active antenna module for low-cost electronically scanned phased arrays. IEEE Trans. Microw. Theory Tech. **56**(10), 2286–2292 (2008)

[KP06] N. Kingsley, J. Papapolymerou, Organic "wafer-scale" packaged miniature 4-bit rf mems phase shifter. IEEE Trans. Microw. Theory Tech. **54**(3), 1229–1236 (2006)

[KPP08] N. Kingsley, G.E. Ponchak, J. Papapolymerou, Reconfigurable rf mems phased array antenna integrated within a liquid crystal polymer (lcp) system-on-package. IEEE Trans. Antennas Propag. **56**(1), 108–118 (2008)

[Kos00] M.A. Kossel, *Active Microwave Tagging Systems*. Ph.D. thesis, Swiss Federal Institute Of Technology Zurich (2000)

[Kry+11] V.A. Krylov et al., Supervised high-resolution dual-polarization sar image classification by finite mixtures and copulas. IEEE J. Sel. Topics Signal Process. **5**(3), 554–566 (2011)

[Li+10] Yue Li et al., Polarization reconfigurable slot antenna with a novel compact cpw-to-slotline transition for wlan application. IEEE Antennas Wirel. Propag. Lett. **9**, 252–255 (2010)

[Mou+08] P. Mousavi et al., A low-cost ultra low profile phased array system for mobile satellite reception using zero-knowledge beamforming algorithm. IEEE Trans. Antennas Propag. **56**(12), 3667–3679 (2008)

[Mue07] S. Mueller, *Grundlegende Untersuchungen steuerbarer passiver Flüssigkristall-Komponenten für die Mikrowellentechnik*. Ph.D. thesis, Technische Universitaet Darmstadt (2007)

[Nov+97] L.M. Novak et al., Effects of polarization and resolution on sar atr. IEEE Trans. Aerosp. Electron. Syst. **33**(1), 102–116 (1997)

[PC10] J. Perruisseau-Carrier, Dual-polarized and polarization-flexible reflective cells with dynamic phase control. IEEE Trans. Antennas Propag. **58**(5), 1494–1502 (2010)

[PCD10] D. Piazza, M. Capacchione, M. D'Amico, Crlh leaky wave antenna with tunable polarization. in *Proceedings of the Fourth European Conference Antennas and Propagation (EuCAP)* (2010), pp. 1–5

[PD06] D. Piazza, K.R. Dandekar, Reconfigurable antenna solution for mimo-ofdm systems. Electron. Lett. **42**(8), 446–447 (2006)

[PZ02] D. Parker, D.C. Zimmermann, Phased arrays-part 1: theory and architectures. IEEE Trans. Microw. Theory Tech. **50**(3), 678–687 (2002)

[Qin+10] Pei-Yuan Qin et al., Polarization reconfigurable u-slot patch antenna. IEEE Trans. Antennas Propag. **58**(10), 3383–3388 (2010)

[Rom+00] R.R. Romanofsky et al., K-band phased array antennas based on ba0.60sr0.40 tio3 thin-film phase shifters. IEEE Trans. Microw. Theory Tech. **48**(12), 2504–2510 (2000)

[SHK09] B. Sanadgol, S. Holzwarth, J. Kassner, 30 ghz liquid crystal phased array. in *Loughborough Antennas and Propagation Conference LAPC 2009* (2009)

[Saz+11b] M. Sazegar et al., Low-cost phased-array antenna using compact tunable phase shifters based on ferroelectric ceramics. IEEE Trans. Microw. Theory Tech. **59**, 1265–1273 (2011)

[Sch+81] D. Schaubert et al., Microstrip antennas with frequency agility and polarization diversity. IEEE Trans. Antennas Propag. **29**(1), 118–123 (1981)

[TTO91] S. Taira, M. Tanaka, S. Ohmori, High gain airborne antenna for satellite communications. IEEE Trans. Aerosp. Electron. Syst. **27**(2), 354–360 (1991)

[Top+08] K. Topalli et al., A monolithic phased array using 3-bit distributed rf mems phase shifters. IEEE Trans. Microw. Theory Tech. **56**(2), 270–277 (2008)

[Tos08] A.G. Toshev, Multipanel concept for wide-angle scanning of phased array antennas. IEEE Trans. Antennas Propag. **56**(10), 3330–3333 (2008)

[WR06] J.-F. Wu, J.-S. Row, Broadband circularly-polarised microstrip antenna with switchable polarisation sense. Electron. Lett. **42**(24), 1374–1375 (2006)

[WR09] C.R. White, G.M. Rebeiz, Single—and dual-polarized tunable slot-ring antennas. IEEE Trans. Antennas Propag. **57**(1), 19–26 (2009)

[ZYG04] Shun-Shi Zhong, Xue-Xia Yang, Shi-Chang Gao, Polarization-agile microstrip antenna array using a single phase-shift circuit. IEEE Transactions on Antennas and Propagation **52**(1), 84–87 (2004)

[Zhe11] Y. Zheng, *Tunable Multiband Ferroelectric Devices for Reconfigurable RF-Frontends*. Ph.D. thesis, Technische Universitaet, Darmstadt (2011)

Chapter 2
Liquid Crystal Material for Microwave Applications

Liquid crystal (LC) molecules feature an additional state of matter, which exits between a liquid and a solid state. In this state, the LC molecules can flow even though they have orientational order. The molecules can be formed in plenty of different forms depending on how the constituents are bonded. These forms are mainly in a rod- or disk-like shape. The shape anisotropy results in different physical phenomena such as the liquid crystal mesophase (form Greek, mesos = middle), anisotropies in the reflective index, in the magnetic susceptibility and in the dielectric properties. These fundamental properties of rod-like shape nematic liquid crystals are presented in the first section in accordance with the scope of this work. In the second section, it is aimed to give reader insights into LC based planar microwave devices. Hence, required technology and working principle of a variable delay line are presented.

2.1 Fundamentals of Liquid Crystal Materials

The most common states of matter on Earth are solid, liquid and gas depending on the temperature and pressure. In the solid state, the molecules are strongly bonded by intermolecular forces either in a regular order, so called crystalline solid, or in an irregular order. In the crystalline solid state, the molecules can only vibrate at their well defined positions (positional order) and have a tendency to align in a specific direction (orientational order). As the temperature of the matter increases to a certain *melting point* T_m, the solid state transforms into a liquid state, where the molecules can move freely and feature no more positional and orientational orders.

The strengths of the intermolecular forces in the LC materials are not homogenous in all directions because the material is formed by anisotropic molecules. When the temperature of the solid LC material is increased, the molecules vibrates excessively and breakdown the weak intermolecular bonds, leading to a drop off in the positional order [DEM+98]. This middle state is defined as the liquid crystal state, where the

O. H. Karabey, *Electronic Beam Steering and Polarization Agile Planar Antennas in Liquid Crystal Technology*, Springer Theses, DOI: 10.1007/978-3-319-01424-1_2,

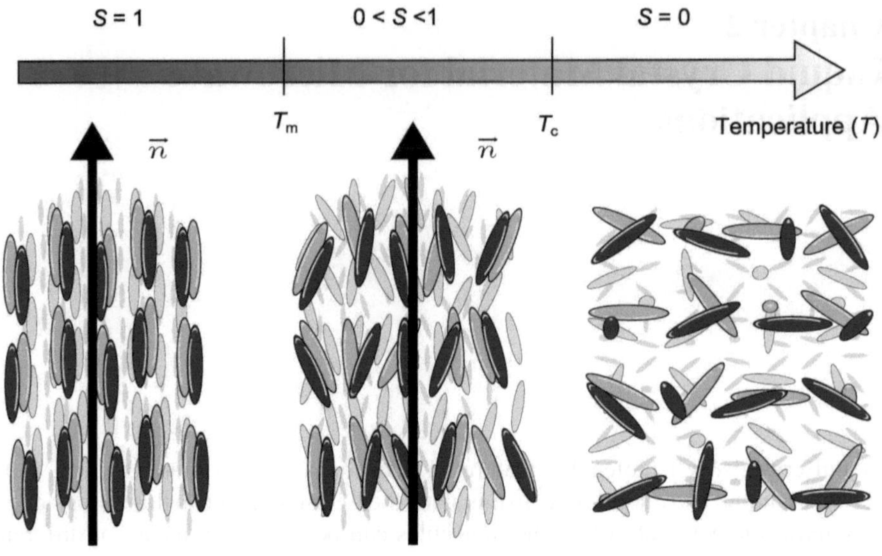

Fig. 2.1 Schematic of a nematic liquid crystal material depending on the temperature. **a** Crystalline solid state with perfect alignment with the order parameter $S = 1$. Director $\vec{n}$ is parallel to the molecular *long axis*. **b** Nematic phase, where $\vec{n}$ indicates the time-average directions of the molecules. **c** Liquid phase with the lowest molecular ordering like $S = 0$

molecules still exhibit orientational order like of a solid crystalline and the material can additionally flow like a liquid [YW06].

Figure 2.1 shows the schematic representation of a nematic liquid crystal. In an LC bulk, the molecules are aligned parallel to their long axes because of the shape anisotropy. In the macroscopic scale, the time-averaged direction of the molecules is denoted by a director $\vec{n}$, which is along the long axes of the molecules. Such a preferred direction is not defined for the short axes since the nematic liquid crystals are usually uniaxial [YW06]. In the both solid and nematic phases, a director can be assigned whereas the level of molecular ordering is different. This is quantitatively defined by an order parameter S [Goe09]. At temperature T_c, so called the *clearing point*, the transition into the liquid phase occurs and the opacity clears. 5CB[1] is a well-known commercially available LC and frequently used in the LC display applications. The clearing point of this material is about 38 °C [Goe09] (also see Appendix A.2.1), which is low for an application. Recently, both experimental [Goe09] and commercial LC mixtures (GT3 series from Merck KGaA, Germany), which are synthesized especially for microwave applications, show an increased nematic range, i.e. from $T_m = -20\,°\mathrm{C}$ to $T_c = 100\,°\mathrm{C}$ for GT3-23001.

Arrangement of the LC molecules in a bulk is determine quantitatively by using the order parameter S [GP93]. Figure 2.2 shows a snapshot of a single rod-like shape

[1] Distributed under the trade name K15 from Merck KGaA, Germany.

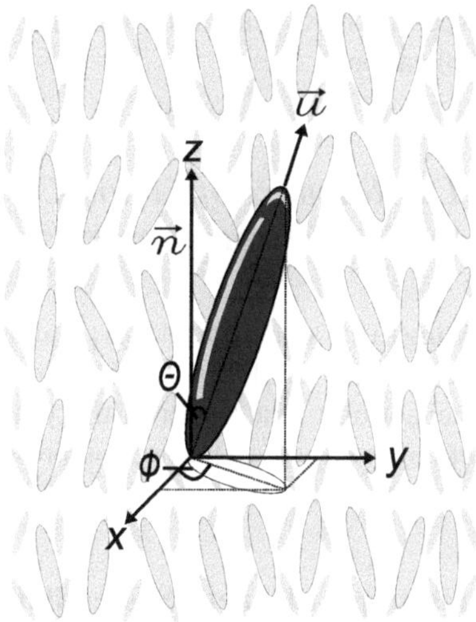

Fig. 2.2 The coordinate system for a single liquid crystal molecule. The material is in the nematic phase. The director $\vec{n}$ is aligned along the z *axis*. The molecule *long axis* is parallel to the unit vector $\vec{u}$ at the given observation time

LC molecule. The single molecule points in $\vec{u}$ direction at the moment of observation and the director is parallel to the z axis. This means that the average direction of the molecules is in the z direction or the molecules spend more time when pointing in the z direction than other directions [CH97].

The probability of directing the unit vector $\vec{u}$ within the solid angle $\mathrm{d}\Omega$ that is $\mathrm{d}\Omega = \sin\theta \mathrm{d}\theta \mathrm{d}\phi$ is given by $f(\theta, \phi)\mathrm{d}\Omega$ [YW06]. Here, $f(\theta, \phi)$ is an orientational distribution function. Since the nematic LCs are uniaxial, there is no preferred direction along ϕ and therefore, $f(\theta, \phi)\mathrm{d}\Omega$ can be rewritten as $f(\theta)\mathrm{d}\Omega$ with $\mathrm{d}\Omega = \sin\theta \mathrm{d}\theta$. The order parameter S can be determined by averaging $f(\theta)\mathrm{d}\Omega$ over the time and over the space. A commonly used expression for S is obtained by using the average of the 2*nd* order Legendre polynomial [CH97, YW06], which is given by

$$S = \langle P_2(\cos\theta)\rangle = \left\langle \frac{3}{2}\cos^2\theta - \frac{1}{2}\right\rangle = \frac{\int_0^\pi \left(\frac{3}{2}\cos^2\theta - \frac{1}{2}\right) f(\theta)\, \sin\theta \mathrm{d}\theta}{\int_0^\pi f(\theta)\, \sin\theta \mathrm{d}\theta}. \tag{2.1}$$

When the molecules are perfectly aligned, $f(\theta) = \delta(\theta)$ and the order parameter $S = 1$ [CH97] according to Eq. (2.1). For a completely unordered case, on the other hand, the molecules direct to all directions with an equal probability. Thus, $f(\theta)$ is constant, resulting to $S = 0$ [YW06]. In the nematic phase, the order parameter is in the range of 0–1 depending on the temperature. To calculate S in the nematic phase, $f(\theta)$ can be described in details by using *Maier-Saupe* or *Landau-de Gennes theory* [GP93, CH97, YW06]. The first considers the temperature dependency of S and the distribution function is determined by the Boltzmann distribution. The second accounts additional effects and enables to fit the respective function to a measurement by use of additional parameters [Goe09]. Further details of these theories are not discussed here since this work focus on the usage of LC materials in microwave applications.

2.1.1 Response of Liquid Crystal Molecules in Electromagnetic Fields

Owing to the shape anisotropy, liquid crystals are diamagnetic and dielectric materials [BC94, CH97, GP93]. They feature either anisotropic magnetic susceptibility $\Delta\chi_m$ or anisotropic relative permittivity $\Delta\varepsilon_r$ if they are exposed to a magnetic or an electric field, respectively. This section presents responses of the LC materials to an applied electromagnetic field.

Anisotropic Magnetic Susceptibility $\Delta\chi_m$

The liquid crystal molecules have magnetic polarizability κ when they are exposed to a magnetic field $\vec{H}$. The magnetic polarizability κ of the molecules differs depending on the angle between $\vec{H}$ and $\vec{u}$ [YW06]. It can be decomposed into $\kappa_{\parallel}$ and $\kappa_{\perp}$, where the terms $\parallel$ and $\perp$ stand for the directions parallel and perpendicular to the molecule long axis, respectively. A diagonal tensor of the magnetic polarizability $\overleftrightarrow{\tilde{\kappa}}$, which is

$$\overleftrightarrow{\tilde{\kappa}} = \begin{pmatrix} \kappa_{\perp} & 0 & 0 \\ 0 & \kappa_{\perp} & 0 \\ 0 & 0 & \kappa_{\parallel} \end{pmatrix}, \qquad (2.2)$$

can be written for a uniaxial LC material along $\vec{u}$ [Goe09]. The unit vector $\vec{u}$ is given in the spherical coordinate system with $\vec{u} = \vec{u}_\theta + \vec{u}_\phi + \vec{u}_r$. $\overleftrightarrow{\tilde{\kappa}}$ is transformed into the cartesian coordinate system xyz in Fig. 2.2 by using a rotation matrix R [CH97]. The resultant magnetic polarizability tensor $\overleftrightarrow{\kappa}$ is

$$\overleftrightarrow{\kappa} = R\,\overleftrightarrow{\tilde{\kappa}}\,R^{-1} \qquad (2.3)$$

with

$$R = \begin{pmatrix} \cos\theta\cos\phi & -\sin\phi & \sin\theta\cos\phi \\ \cos\theta\sin\phi & \cos\phi & \sin\theta\sin\phi \\ -\sin\theta & 0 & \cos\theta \end{pmatrix}. \qquad (2.4)$$

The magnetic susceptibility χ_m is a macroscopic property. It can be obtined by multipling the magnetic polarizability κ by the molecular number density N [YW06, Goe09]. The magnetic susceptibility is written as $\overleftrightarrow{\chi_m} = \mathrm{N}\overleftrightarrow{\kappa}$ in the tensorial form. The derivations up to here are carried out only for one molecule. As described in Sect. 2.1, the averaged magnetic susceptibility of an LC bulk can be obtained by using $\langle\rangle$ operator element by element. By calculating $\langle\overleftrightarrow{\chi_m}\rangle = \mathrm{N}\langle\overleftrightarrow{\kappa}\rangle$, one obtains [YW06]

$$\langle \overleftrightarrow{\chi_m} \rangle = \begin{pmatrix} \chi_{m,\perp} & 0 & 0 \\ 0 & \chi_{m,\perp} & 0 \\ 0 & 0 & \chi_{m,\parallel} \end{pmatrix}$$
$$= \mathrm{N} \begin{pmatrix} \kappa_\perp + \frac{1}{3}(1-S)\Delta\kappa & 0 & 0 \\ 0 & \kappa_\perp + \frac{1}{3}(1-S)\Delta\kappa & 0 \\ 0 & 0 & \kappa_\perp + \frac{1}{3}(2S+1)\Delta\kappa \end{pmatrix}. \quad (2.5)$$

The anisotropic magnetic susceptibility $\Delta\chi_m$ is

$$\Delta\chi_m = \chi_{m,\parallel} - \chi_{m,\perp} = \mathrm{N}S\Delta\kappa. \quad (2.6)$$

The magnetic susceptibilities of the LC material are negative and close to zero (in the order of 10^{-5} in SI units) [CH97, YW06] and therefore, it is not considered in this work. Nevertheless, as will be given in 2.1.2, the anisotropic magnetic response of the material, although being low, is exploited consciously to control the orientation of the LC molecules.

Anisotropic Relative Permittivity $\Delta\varepsilon_r$

In the presence of an external electrical field $\vec{E}$, a dielectric dipole moment $\vec{p}$ is induced on the LC molecules since the moluces are polarizable. Similar to the magnetic case, the polarizability α_p of the LC molecules differs depending on the angle between $\vec{E}$ and $\vec{u}$. The tensorial material polarizability $\overleftrightarrow{\alpha_p}$ is

$$\overleftrightarrow{\alpha_p} = \begin{pmatrix} \alpha_{p,\perp} & 0 & 0 \\ 0 & \alpha_{p,\perp} & 0 \\ 0 & 0 & \alpha_{p,\parallel} \end{pmatrix} \quad (2.7)$$

according to the molecular axis $\vec{u}$. The dipole moment $\vec{p}$ is actually induced by the local electrical field on a molecule $\vec{E}_{local}$, which is produced by the external field and by the dipole moments of the other molecules [YW06]. Here, $\vec{E}_{local}$ is not equal to the applied external field unlike to the magnetic case because of the strong dipole-dipole interactions between the molecules. The relationship between $\vec{E}_{local}$ and $\vec{E}$ is given as

$$\vec{E}_{local} = \overleftrightarrow{K}\,\vec{E}, \quad (2.8)$$

where $\overleftrightarrow{K}$ is so called the tensorial internal field constant [YW06, Goe09]. It is defined by $\overleftrightarrow{K} = 1/\left(1 - \mathrm{N}\overleftrightarrow{\alpha_p}/3\varepsilon_0\right)$ with the permittivity of free space ε_0 (8.85 x 10^{-12} F/m) and the molecular number density n. The induced dipole moment is obtained by

$$\vec{p} = \overleftrightarrow{\alpha_p}\,\vec{E}_{local} = \overleftrightarrow{\alpha_p}\,\overleftrightarrow{K}\,\vec{E}. \quad (2.9)$$

The corresponding macroscopic polarization $\vec{P}$ is

$$\vec{P} = \mathrm{N}\vec{p} = \mathrm{N}\overleftrightarrow{\underline{\alpha}_p}\,\overleftrightarrow{K}\,\vec{E} \tag{2.10}$$

along to molecular axis $\vec{u}$ is [YW06]. The tensors in Eq. (2.10) can be transformed into the laboratory coordinates xyz (see Fig. 2.2) by using the rotation matrix R. Then, the Eq. (2.10) can be rewritten for the laboratory frame as

$$\begin{aligned}\vec{P} &= \mathrm{N}R\overleftrightarrow{\underline{\alpha}_p}\,\overleftrightarrow{K}\,R^{-1}\vec{E} \\ &= \overleftrightarrow{\underline{\chi}_e}\,\vec{E}.\end{aligned} \tag{2.11}$$

where χ_e and $\overleftrightarrow{\underline{\chi}_e}$ are the electric susceptibility and its tensor, respectively. Averaging $\overleftrightarrow{\underline{\chi}_e}$ by using $\langle\rangle$ operator, one obtains the effective susceptibility tensor as [YW06]

$$\langle\overleftrightarrow{\underline{\chi}_e}\rangle = \frac{\mathrm{N}}{3}\begin{pmatrix} \alpha_\perp K_\perp(2+S) & 0 & 0 \\ +\alpha_\parallel K_\parallel(1-S) & & \\ 0 & \alpha_\perp K_\perp(2+S) & 0 \\ & +\alpha_\parallel K_\parallel(1-S) & \\ 0 & 0 & \alpha_\perp K_\perp(2-2S) \\ & & +\alpha_\parallel K_\parallel(1+2S) \end{pmatrix} \tag{2.12}$$

The dielectric displacement $\vec{D}$ is defined as [CH97]

$$\vec{D} = \varepsilon_0\vec{E} + \vec{P} = (\varepsilon_0 + \overleftrightarrow{\underline{\chi}_e})\vec{E} = \varepsilon_0\overleftrightarrow{\underline{\varepsilon}_r}\,\vec{E}, \tag{2.13}$$

In Eq. (2.13), the complex relative permittivity tensor $\overleftrightarrow{\underline{\varepsilon}_r}$ is defined by $\overleftrightarrow{\underline{\varepsilon}_r} = \overleftrightarrow{I} + \overleftrightarrow{\underline{\chi}_e}/\varepsilon_0$ with the identity matrix tensor $\overleftrightarrow{I}$. By using Eq. (2.12), one can obtain the complex effective perpendicular and parallel relative permittivity

$$\underline{\varepsilon}_{r,\perp} = 1 + \frac{\mathrm{N}}{3\varepsilon_0}[\alpha_\perp K_\perp(2+S) + \alpha_\parallel K_\parallel(1-S)] \tag{2.14}$$

and

$$\underline{\varepsilon}_{r,\parallel} = 1 + \frac{\mathrm{N}}{3\varepsilon_0}[\alpha_\perp K_\perp(2-2S) + \alpha_\parallel K_\parallel(1+2S)], \tag{2.15}$$

respectively. The resultant anisotropic relative permittivity $\Delta\underline{\varepsilon}_r$ is calculated by

$$\Delta\underline{\varepsilon}_r = \underline{\varepsilon}_{r,\parallel} - \underline{\varepsilon}_{r,\perp} = \frac{\mathrm{N}}{\varepsilon_0}(\alpha_\parallel K_\parallel - \alpha_\perp K_\perp) \tag{2.16}$$

and analogously, the anisotropic refractive can be written as

$$\Delta\underline{n} = \underline{n}_\parallel - \underline{n}_\perp = \sqrt{\underline{\varepsilon}_{r,\parallel}} - \sqrt{\underline{\varepsilon}_{r,\perp}} \tag{2.17}$$

since the complex relative permeability is $\underline{\mu}_r \approx 1$. The dielectric anisotropy is proportional to the order parameter and can be either positive or negative. Nevertheless, as will be described in Chap. 3, all LC samples used in this work have positive dielectric anisotropy.

The electrical characteristics of the LC materials are characterized by using two quantities. These are so called the material tunability τ_{LC} and the material efficiency η_{LC}. The first is related to the real part of the complex relative permittivity and the second accounts dielectric losses. The complex relative permittivity can be rewritten as

$$\begin{aligned}\underline{\varepsilon}_{r,i} &= \Re(\underline{\varepsilon}_{r,i}) - j\Im(\underline{\varepsilon}_{r,i}) \\ &= \Re(\underline{\varepsilon}_{r,i})\left(1 - j\frac{\Im(\underline{\varepsilon}_{r,i})}{\Re(\underline{\varepsilon}_{r,i})}\right) \\ &= \varepsilon_{r,i}(1 - j\tan\delta_i),\end{aligned} \tag{2.18}$$

where, i represents the orientation of the LC molecules, i.e $\|$ or $\perp$. In Eq. (2.19), ε_r is the relative permittivity and $\tan\delta$ is the material loss tangent. The corresponding τ_{LC} and η_{LC} are determined by

$$\tau_{\mathrm{LC}} = \frac{\varepsilon_{r,\|} - \varepsilon_{r,\perp}}{\varepsilon_{r,\|}} = \frac{\Delta\varepsilon_r}{\varepsilon_{r,\|}} \tag{2.19}$$

and

$$\eta_{\mathrm{LC}} = \tau_{\mathrm{LC}}/\max(tan\delta), \tag{2.20}$$

respectively. As will be shown in Chap. 3, the $\tan\delta$ experienced by the electromagnetic field is higher if the field is perpendicular to the molecules. This refers that $\tan\delta_\perp > \tan\delta_\|$ at the microwave frequencies [Mue07, Goe09]. Thus, in calculations of η_{LC}, $\tan\delta_\perp$ is utilized because it defines the minimum material efficiency.

2.1.2 Alignment Methods

The polarizability α_p and therefore, the macroscopic relative permittivity $\underline{\varepsilon}_r$ of the LC material depend on the angle between the director $\vec{n}$ and the electric field $\vec{E}$ (see Sect. 2.1.1). In an tunable device, the direction of $\vec{n}$ has to be controlled to exploit the LC material as a tunable dielectric. Possible methods to align the LC molecules are described in this section. Before doing so, it is essential to explain briefly total free energy in an LC bulk and the terms contributing to this energy.

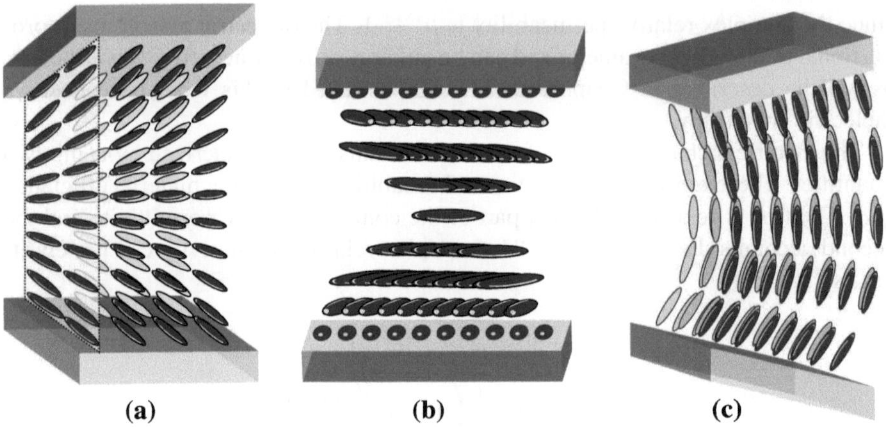

Fig. 2.3 Possible deformations of an oriented LC layer. **a** Splay, **b** twist and **c** bend types of deformations

Total Free Energy in an LC Bulk

The continuum theory [Dem+99, YW06] is a commonly used method for modelling the LC material in a device. Any kind of local or complete change in the direction of $\vec{n}$ increases the energy of the system, which is formulated by the Gibbs free type of energy. The system achieves a equilibrium state when the Gibbs free energy has its minimum at constant temperature and pressure [YW06]. This also means that if the system's energy is increased, the molecular orientation (and $\vec{n}$) changes to find a new equilibrium state, resulting in again a minimum free energy. The factors having impact on the system's energy are elastic deformations, external electromagnetic fields and anchoring effects.

In reality, even the molecules are aligned in a specific direction, the director $\vec{n}$ may vary in the volume depending the position [YW06]. This phenomenon is called the elastic deformation of the director. Three possible deformations can be observed in an oriented LC material. These are splay, twist and bend as shown in Fig. 2.3. The contributions of the deformations to the energy of the system per unit volume can be computed as [CH97, Dem+99]:

$$\begin{aligned} \mathrm{d}f_{\text{elastic}} &= \mathrm{d}f_{\text{splay}} + \mathrm{d}f_{\text{twist}} + \mathrm{d}f_{\text{bend}} \\ &= \frac{1}{2}K_{11}(\nabla \cdot \vec{n})^2 + \frac{1}{2}K_{22}[\vec{n} \cdot (\nabla \times \vec{n})]^2 + \frac{1}{2}K_{33}[\vec{n} \times (\nabla \times \vec{n})]^2, \end{aligned} \tag{2.21}$$

where K_{11}, K_{22} and K_{33} are elastic constants for the splay, twist and bend types of deformations, respectively. These values are characteristics for an LC sample and can be determined by performing measurements as in [Ong91].

Beside elastic distortions, the energy of the system changes in the presence of an external electromagnetic field. In case of an external electric field $\vec{E}$, the energy per unit volume is given by [CH97]

$$f_{\text{electric}} = \frac{1}{2}\vec{D} \cdot \vec{E} \tag{2.22}$$

for a parallel plate varactor, where the LC material is contained between two conducting electrodes as a tunable dielectric. Such a simple device models the working principle of the hardwares realized in this work. An incremental change in the director $\mathrm{d}\vec{n}$ leads to a change in the polarization $\vec{P}$ because of the anisotropic behaviour of the material. One can also write that $\mathrm{d}\vec{P} = \mathrm{d}\vec{D}$ according to (2.13) since the applied voltage across the electrodes and therefore, the applied bias field are constant [CH97]. Thus, the small change in the energy per unit volume is

$$\mathrm{d}f_{\text{electric},1} = \frac{1}{2}\mathrm{d}\vec{P} \cdot E. \tag{2.23}$$

On the other hand, the amount of the charge on the electrodes has to be adopted to keep the applied voltage constant because of the changing polarization. The corresponding small change in the energy per unit volume is [CH97]

$$\mathrm{d}f_{\text{electric},2} = -\mathrm{d}\vec{P} \cdot \vec{E}. \tag{2.24}$$

One can add Eqs. (2.23) and (2.24) to obtain the total change in the energy per unit volume caused by the electric field. This is negative and equal to

$$\mathrm{d}f_{\text{electric}} = \mathrm{d}f_{\text{electric},1} + \mathrm{d}f_{\text{electric},2} = -\frac{1}{2}\mathrm{d}\vec{P} \cdot \vec{E}. \tag{2.25}$$

Equation (2.25) is given in terms of the relative permittivity and the director as [Goe09]

$$\mathrm{d}f_{\text{electric}} = -\frac{1}{2}\underline{\varepsilon}_0(\langle \vec{E}^2 \rangle \varepsilon_{r,\perp} + \Delta\underline{\varepsilon}_r \langle (\vec{n} \cdot \vec{E})^2 \rangle). \tag{2.26}$$

Similarly, the change in the energy per unit volume caused by the magnetic field $\vec{H}$ is [Mue07]

$$\mathrm{d}f_{\text{magnetic}} = -\frac{1}{2}\mu_0(\langle \vec{H}^2 \rangle \underline{\mu}_{r,\perp} + \Delta\underline{\mu}_r \langle (\vec{n} \cdot \vec{H})^2 \rangle) \tag{2.27}$$

with μ_0 is permeability of free space ($4\pi \mathrm{x} 10^{-7}$ N/A^2) and μ_r is relative permeability.

Another factor affecting the system's energy is anchoring forces because of the boundaries. In many liquid crystal devices, the viscous LC material is contained between two substrates. These substrates are coated with an alignment layer. When the anisotropic LC material has a contact with the anisotropic surface (because of the alignment layer) interaction energy occurs. This is called the anchoring energy. According to *Rapini-Papoular Model*, this energy density is determined as [YW06]

$$\mathrm{d}f_{\text{surface}} = \frac{1}{2}W_p \sin^2(\theta - \theta_0) + \frac{1}{2}W_a \sin^2(\phi - \phi_0). \tag{2.28}$$

In Eq. (2.28), W_p and W_a are polar and azimuthal anchoring strengths, respectively; (θ, ϕ) specifiy the director $\vec{n}$ and (θ_0, ϕ_0) specifiy the preferred alignment direction for the polar and azimuthal angles, respectively.

The total free energy W_f of the system can be written as

$$W_f = \int\limits_{\text{Volume}} \mathrm{d}f = \int\limits_{\text{Volume}} (\mathrm{d}f_{\text{elastic}} + \mathrm{d}f_{\text{electric}} + \mathrm{d}f_{\text{magnetic}} + \mathrm{d}f_{\text{surface}}). \tag{2.29}$$

Alignment by Surface Anchoring

The preferred alignment direction defined by the alignment layer can be either in parallel (homogenous alignment) or perpendicular (homeotropic alignment) to the surface [YW06]. The alignment principal can be explained in the light of the energy considerations. The system's total energy is proportional to the anchoring surface energy f_{surface} (see Eq. (2.29)) and therewith f_{surface} has to be minimized for an equilibrium state of the molecules. According to Eq. (2.28), f_{surface} is minimized when $\sin(\theta - \theta_0)$ and $\sin(\phi - \phi_0)$ terms are zero for an ideal case. This results in $\theta = \theta_0$ and $\phi = \phi_0$, which means that $\vec{n}$ (specified by θ and ϕ) is aligned parallel to the preferred direction (specified by θ_0 and ϕ_0).

In this work, a homogenous alignment layer is utilized that is made of Nylon 6. It aligns the molecules parallel to the surfaces at a zero biasing case, which refers that no external field is applied. The alignment layer is spin coated, cured and then mechanically rubbed [Goe09] to have microscopic grooves on the surface. The coating parameters of the alignment layer are crucial for the device performance. A thick alignment layer in the order of a few mm affects the radio frequency (RF) performance of the device. On the other hand, a thin alignment layer results in light and nonuniform grooves having a low anchoring force. Several different coating parameters for the alignment layer are tested and these results are given in Appendix A.1.2.

Alignment by Electromagnetic Fields

In the precense of an external electric $\vec{E}$ or magnetic $\vec{H}$ field, the liquid crystal molecules are aligned to minimize the total free energy.

The energy contributed by an external electrical field f_{electric} is negative (see Eq. (2.26)) and therefore, increasing its absolute value decreases the total free energy. LC materials used for the microwave applications feature a positive anisotropy [Mue07, Goe09] that is $\Delta\varepsilon_r = \varepsilon_{r,\parallel} - \varepsilon_{r,\perp} > 0$. Hence, $|f_{\text{electric}}|$ is maximized when the director is aligned parallel to the applied field. This can be seen in the second term of Eq. (2.26).

A similar alignment can be accomplished by using the magnetic fields (see Eq. (2.27)). However, this method is rarely used, especially for LC devices, because of the heavy and bulky magnets. Nevertheless, for the experimental setups, i.e. for characterization of the material parameters, the molecules can be aligned parallel to an applied magnetic field.

2.2 Fundamentals of Tunable Inverted Microstrip Lines

This section aims to give reader insights into fundamentals of LC based microwave devices. For this purpose, an LC based planar variable delay line is presented including required technology and its working principal. Here the term planar device refers a microwave device, which is implemented by using a printed transmission line such as microstrip line, slotline or coplanar waveguide. In this context, the variable delay line is chosen in inverted microstrip line (IMSL) topology since it is a well-known application of LCs for microwaves [KFN02, Mue07].

2.2.1 Technology

Figure 2.4 shows a schematic of the fabrication process of the IMSL. Similar to liquid crystal displays (LCDs), glass substrates are preferred for microwave device realization. The reasons are that the glass is mechanically stable, which is essential to achieve a uniform LC layer thickness, and it is transparent, so different glass substrates can be aligned and stacked on top of each other. The manufactures delivers dielectric properties only for low frequency, i.e. for 1 MHZ. The characterization of glass substrates are given in Sect. 3.2 for microwave frequencies (see also Appendix A.2.2).

Electrodes of the IMSL are implemented on the glass substrates by using photolithography processes. Technological details can be found in Appendix A.1.1. A seed layer, which is made of Nickel–Chromium (Ni–Cr) and gold (Au) metals, is evaporated on the glass substrate. The Ni-Cr layer is utilized as an adhesive layer between the substrate and Au layer. Typical thicknesses are 15 nm for Ni–Cr layer and 60 nm for the Au layer. A positive photoresist is spin coated on top of the seed layer. The photoresist is exposed with UV light and subsequently developed by using a convenient developer (see Appendix A.1.1). The signal and ground electrodes of the IMSL are formed inside the well-defined valleys by using electroplating of the gold metal. The thickness of the plated gold is specified to be 3 times of the skin depth at the device's operating frequency. This value is chosen as a compromise between a

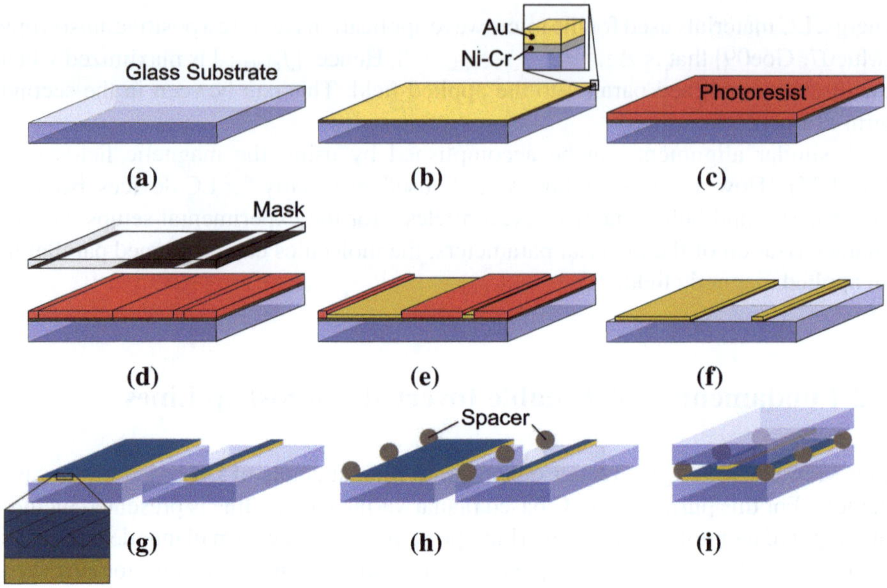

Fig. 2.4 A schematic overview of the fabrication processes of LC based inverted microstrip line. **a** Top of a glass substrate is covered by **b** nickel–chromium (Ni–Cr)/gold (Au) seed layer by using thermal evaporation. **c** A positive photoresist is spin coated, **d** exposed with UV light and developed. **e** The electrodes of the IMSL are formed inside the valleys by electroplating of the gold electrode. **f** The photoresist is removed and the seed layer is etched. **g** The substrate is diced into two pieces and each substrate is coated with an alignment layer. **h** Spacers are formed on the top of the substrates. **i** Finally, the two substrates are aligned and bonded by UV curable adhesive

lower electrode thickness, causing a high metallic loss, and a higher electrode thickness, raising the fabrication costs. During the plating, two types of marker are also formed on the glass substrate whereas they are not shown in Fig. 2.4. These markers are used in the next steps for cutting and aligning purposes. After the plating, the photoresist is stripped and the seed layer is etched first by using a Au etchant and then by using a Ni–Cr etchant. The substrate is diced into two pieces by means of the cutting markers. Each piece is coated with an alignment layer made of Nylon 6 and rubbed mechanically to form grooves on the surfaces (see Appendix A.1.2). The substrates are covered with spherical glass spacers having 100 mm diameter. The spacers are commercially available from Sekisui Chemical Co., Ltd. In LC displays, similar spacer are used with lower dimensions. The substrate carrying the signal electrode is inverted and aligned on top of the ground electrode by using the complementary alignment marks. The aligned substrates are bonded with UV curable adhesive that is Ablestik A4086T. The resultant structure is that the signal electrode is on the bottom side of the front glass substrate and the ground electrode is on the top side of the back glass substrate. In between, a cavity with 100 mm height is formed because of the spacers. The LC material is filled into the cavity by means of adhesion forces and subsequently. Finally, the structure is sealed either by using UV sensitive glue or any kind of nonconductive two-component glues.

2.2.2 Working Principle

Cross-sectional views of the IMSL is shown in Fig. 2.5. Owing to the alignment layer, the entire LC bulk is oriented parallel to the surfaces initially. The director $\vec{n}$ is perpendicular to the RF field distribution of the IMSL. The electric field, propagating in the LC layer, experiences the complex perpendicular relative permittivity $\underline{\varepsilon}_{r,\perp}$, which is along the molecule's short axis. This is given as

$$\underline{\varepsilon}_{r,\perp} = \varepsilon_{r,\perp}(1 - j \tan \delta_{\perp}). \tag{2.30}$$

according to Eq. (2.18). Indeed, the main part of the field propagates in the LC layer and only some in the front glass, which is called as fringing field. Hence, the propagating wave experiences an *effective permittivity* depending on the device geometry and the electrical parameters of the glass substrates [Bal05] . The challenge is to force the field to propagate only through the LC layer by reducing the fringing field. This enhances the tuning efficiency of the IMSL. Methods reducing the fringing field are presented in Sect. 4.1.1. Nonetheless, the corresponding complex perpendicular effective permittivity $\underline{\varepsilon}_{r,\text{eff},\perp}$ is mainly defined by $\underline{\varepsilon}_{r,\perp}$.

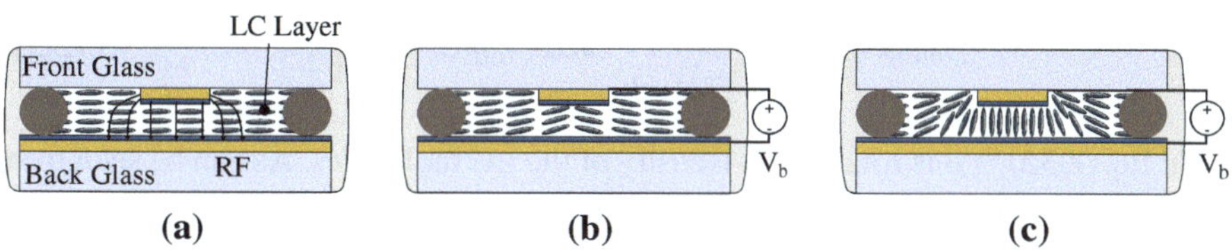

Fig. 2.5 Cross-sectional views of the variable IMSL. **a** The LC molecules are parallel to the surface ($\varepsilon_{r,\text{eff}} = \varepsilon_{r,\text{eff},\perp}$) and perpendicular to the radio frequency (RF) field. **b** The molecus are aligned as equilibrium between the applied field and the anchoring force of the alignment layer ($V_b > V_{\text{th}}$ and $\varepsilon_{r,\text{eff},\perp} < \varepsilon_{r,\text{eff}} < \varepsilon_{r,\text{eff},\parallel}$). **c** The LC molecules are oriented parallel to the electric field of the IMSL ($V_b > V_{\text{sat}}$ and $\varepsilon_{r,\text{eff}} = \varepsilon_{r,\text{eff},\parallel}$)

The director can be realigned by applying an external bias voltage V_b across the signal and the ground electrodes (see Sect. 2.1.2). However, the surface anchoring resists to the realignment caused by the electrical field [YW06]. Only after a threshold voltage is exceeded the molecules are realigned as equilibrium between the applied field and the anchoring force of the alignment layer. This threshold phenomenon is well-known as *Fréedericksz transition* [CH97, YW06]. For the IMSL type LC devices, where the splay deformation is dominant, the required threshold voltage V_{th} is [Goe09]

$$V_{\text{th}} = \pi\sqrt{\frac{K_{11}}{\varepsilon_0 \Delta\varepsilon_r}}. \tag{2.31}$$

Increasing the applied bias voltage dominates the alignment by the external field and therefore, the angle between $\vec{n}$ and the electric field of the IMSL decreases. Eventually, a saturation state is achieved [Goe09], where the molecules are parallel to the electric field of the IMSL. In this case, $\varepsilon_{r,\parallel}$ and $\tan\delta_{\parallel}$ (see Eq. (2.30)) are effective and define the complex parallel effective permittivity $\underline{\varepsilon}_{r,\text{eff},\parallel}$ (see Appendix A.3.3).

Response Times

Time interval required for the transition from an equilibrium state into another is defined as *response time* of the LC devices[Mar08, Goe09]. The response time depends on the height of the LC layer h_{LC}, alignment method and the LC material itself. In this work, two different response times are frequently used. These are given for two extreme alignments of the director, which are either from perpendicular to parallel or parallel to perpendicular alignments with respect to the electric field of the IMSL. For the first one ($\perp$ to $\parallel$), a bias voltage V_b is applied and the time interval is denoted as t_{on}. For the second one ($\parallel$ to $\perp$), the applied voltage is released and the desired molecular orientation is forced by the alignment layer. This is denoted as t_{off}. The formulas for t_{on} and t_{off} are [Mar08]

$$t_{\text{on}} \propto \frac{\gamma_{\text{LC}} h_{\text{LC}}^2}{\Delta\varepsilon_r (V_b^2 - V_{\text{th}}^2)} \quad \text{and} \quad t_{\text{off}} \propto \frac{\gamma_{\text{LC}}\, h_{\text{LC}}^2}{\Delta\varepsilon_r V_{\text{th}}^2}. \tag{2.32}$$

In Eq. (2.32) γ_{LC} is rotational viscosity of the LC molecules. As can be seen from Eq. (2.32), the response times are linearly dependent to the material parameter and quadratic dependent to the LC layer thickness. Hence, the challenge is to propose novel device topologies allowing to lower LC layer thickness to reduce the response times of the LC devices.

Electrical Properties of the IMSL

The electrical length ϕ_ℓ of the variable IMSL can be written as [Che94]

$$\begin{aligned} j\phi_\ell &= \underline{\gamma}\ell_{\text{physical}} \\ &= (\alpha + j\beta)\ell_{\text{physical}} \\ &\approx \frac{j2\pi f_0}{c_0}\sqrt{\varepsilon_{r,\text{eff}}}\left(1 - j\frac{\tan\delta_{\text{eff}}}{2} + \frac{(\tan\delta_{\text{eff}})^2}{8}\right)\ell_{\text{physical}} \end{aligned} \tag{2.33}$$

since the effective relative permeability of LC is $\underline{\mu}_{r,\text{eff}} \approx 1$ and LC is a low-loss dielectric that is $\Im(\underline{\varepsilon}_{r,\text{eff}}) \ll \Re(\underline{\varepsilon}_{r,\text{eff}})$. In Eq. (2.33), $\underline{\gamma}$ is complex propagation constant of the IMSL, including the attenuation α and propagation β constants, ℓ_{physical} is physical length of the IMSL, f_0 is operating frequency and c_0 is speed of light in vacuum ($\approx$ 3x10^8 m/s). The resultant phase shift ϕ_β can be written as

$$\phi_\beta = \Im(j\phi_\ell) = \beta\ell_{\text{physical}} \approx \frac{360°}{c_0} f_0 \ell_{\text{physical}} \sqrt{\varepsilon_{r,\text{eff}}} \tag{2.34}$$

since $(\tan\delta_{\text{eff}})^2/8 \ll 1$ in Eq. (2.33) for the low-loss LC material. It should be noted that $\varepsilon_{r,\text{eff}}$ is a voltage tunable quantity, leading to a voltage dependent phase shift. Therefore, a differential phase shift $\Delta\phi_\beta$ can be defined depending on the applied voltage V_b. This is given by, $\Delta\phi_\beta$ is

$$\begin{aligned} \Delta\phi_\beta &= \phi_\beta(V_b > V_{\text{th}}) - \phi_\beta(V_b = 0) \\ &= \frac{360°}{c_0} f_0 \ell_{\text{physical}} \sqrt{\varepsilon_{r,\text{eff}}(V_b > V_{\text{th}}) - \varepsilon_{r,\text{eff}}(V_b = 0)}. \end{aligned} \tag{2.35}$$

A maximum differential phase shift $\Delta\phi_{\beta,max}$ is achieved when $V_b \geq V_{\text{Sat}}$, where V_{Sat} is so called the saturation voltage. Meantime, the voltage dependent insertion loss IL of the IMSL is calculated by [Mue07]

$$\text{IL}(V_b) \approx 4.343 \tan\delta_{\text{eff}} \frac{2\pi f_0}{c_0} \ell_{\text{physical}} \sqrt{\varepsilon_{r,\text{eff}}(V_b)} \quad [\text{dB}]. \tag{2.36}$$

As can be seen in Eq. (2.35) and (2.36), both differential phase shift and the insertion loss are proportional to ℓ_{physical}. The common parameter for quantifying the RF performance of a variable delay line is a frequency dependent variable delay line efficiency (or phase shifter efficiency) η_{VDL}. It is defined by the ratio of the maximum differential phase shift and the highest insertion loss in all tuning states. This can be formulated by

$$\eta_{\text{VDL}} = \Delta\phi_{\beta,\text{max}}/\text{IL}_{\text{max}} \quad [°/\text{dB}] \tag{2.37}$$

In general, the aim is to achieve the highest possible differential phase shift accompanied by the lowest insertion loss which leads to a high η_{VDL}.

2.3 Summary

This section presented the fundamental material properties of LCs and their usage in a variable delay line (VDL) as tunable dielectric. In Sect. 2.1, it was explained that the LC molecules are uniaxial anisotropic materials, which is originated from its rod-like shape. This results in orientation dependent responses of the molecules to several physical phenomena. In particular, their special responses to the electromagnetic fields were descried. In case of an electrical field only, such a response refers that the LC molecules feature different polarization depending on how the field is incident on the molecules. This results in different macroscopic permittivities. Given that, the orientation between the electric field and the molecules has to be controlled to exploit LCs as tunable dielectrics. For this manner, system's total free energy was described. The factors contributing to this energy were the elastic forces, external electromagnetic fields and surface anchoring forces. In the second part, the LC based inverted microstrip delay line was introduced. Especially, the technology required and working principle of the device were presented. In an application, the LC molecules are aligned by using two methods simultaneously. First, an alignment layer anchors the molecules parallel to the surfaces. Second, the molecules are oriented parallel to the applied biasing field, which is perpendicular to the surfaces. In relation to the first part, the both surface anchoring and bias field increase the system's total free energy initially. Then the molecules reach its equilibrium state gradually while minimizing the total free energy. Therefore, a continuous tunability of the LC material is possible depending on the applied bias voltage. Finally, an essential parameter was defined to measure the quality of different delay lines. This was the delay line efficiency η_{VDL}, which was determined by the ratio of the maximum differential phase shift and the highest insertion loss.

References

[DEM+98] D. Demus et al., *Low Molecular Weight Liquid Crystals I*, vol. 2A (Verlag GmbH, Handbook of Liquid Crystals (WILEY-VCH, 1998)

[YW06] D.-E. Yang, S.-T. Wu, Fundamentals of Liquid Crystal Devices (Wiley, Chichester, 2006)

[Goe09] F. Goelden, Liquid Crystal Based Microwave Components with Fast Response Times: Material, Technology, Power Handling Capability, PhD thesis, Technische Universitaet Darmstadt, 2009.

[GP93] P.G. de Genne, J. Prost, *The Physics of Liquid Crystals* (Oxford Science Publications, Oxford, 1993)

[CH97] P.J. Collings, M. Hird, *Introduction to Liquid Crystals: Chemistry and Physics* (Taylor and Francis Ltd., London, 1997)

[BC94] L.M. Blinov, V.G. Chigrinov, in Electrooptic Effects in Liquid Crystal Materials (Partially Ordered Systems), ed. by L. Lam, D. Langevin (Springer, New York, 1994).

[Mue07] S. Mueller, Grundlegende Untersuchungen steuerbarer passiver Flüssigkristall-Komponenten für die Mikrowellentechnik, PhD thesis, Technische Universitaet Darmstadt, 2007.

[Dem+99] D. Demus et al., *Physical Properties of Liquid Crystals* (WILEY-VCH, Verlag GmbH, 1999)

[Ong91] H.L. Ong, Measurement of nematic liquid crystal splay and bend elastic constants with obliquely incident light. J. Appl. Phys. **70**, 2023–2030 (1991)

[KFN02] T. Kuki, H. Fujikake, T. Nomoto, Microwave variable delay line using dual-frequency switching-mode liquid crystal. IEEE Trans. Microw. Theory Tech. **50**, 2604–2609 (2002)

[Bal05] C.A. Balanis, *Antenna Theory: Analysis and Design*, 3rd edn. (Wiley, New York, 2005)
[Mar08] R. Marin, Investigations on Liquid Crystal Reconfigurable Unit Cells for mm-Wave Reflectarrays, PhD thesis, Technische Universitaet Darmstadt, 2008.
[Che94] D.K. Cheng, *Fundamentals of Engineering Electromagnetics* (Addison-Wesley Publishing Company, New York, 1994)

Chapter 3
Microwave Material Characterization

In a design of a tunable device, precise knowledge of material parameters is essential to know; indeed it is the first data to be provided, to define feature sizes in a layout. In accordance with the scope of this work, the materials refer the tunable liquid crystals (LCs) and glass substrates. This chapter presents methods for extracting the material parameters, which are relative permittivity ε_r and loss tangent $\tan\delta$, in microwave frequencies.

In the first section, the LC materials are characterized over a wide frequency and temperature ranges. For this purpose, the *coaxial line method* is utilized. An extraction algorithm is described, which combines benefits of previously proposed analytical and numerical solutions. This algorithm enables also to extract effects of metallic losses to determine dielectric losses precisely. Additionally, the accuracy of the extracted results are discussed by considering a possible error caused by the setup. In the second part, the glass substrates are characterized by using two methods. The first method is a resonator method, namely *split-cylinder-cavity method.* It enables to characterize the samples at 20 GHz. The second one is a broadband *on-wafer characterization method.* A coplanar waveguide is implemented in the glass substrates. Then, the material parameters are extracted by using the measured scattering parameters.

3.1 Broadband Microwave Characterization of Liquid Crystals

Coaxial line method is frequently utilized for broadband material characterization as given in [Che+04, BJ+93, Mue07]. In a coaxial line, an inner conductor is surrounded by a solid dielectric material and both are then shielded by an outer conductor [Poz11]. In this work, an artificial coaxial line is required since the characterization of liquid materials is aimed. This setup is presented in the following part.

O. H. Karabey, *Electronic Beam Steering and Polarization Agile Planar Antennas in Liquid Crystal Technology*, Springer Theses, DOI: 10.1007/978-3-319-01424-1_3,

3.1.1 Coaxial Line Characterization Method

Measurement Setup

A material under test (MUT) has to be filled between the inner and outer conductors of a coaxial line. The unknown material parameters of the MUT affect the measurable electrical properties of the coaxial line such as its scattering parameters. These parameters are then utilized for the material characterization. Similar to [Mue07], a temperature-controlled artificial coaxial line is built enabling to fill a liquid material as a dielectric of the transmission line. A cross-sectional schematic of the setup is shown in Fig. 3.1.

The setup consists of four main parts: a semi rigid coaxial cable, split blocks, temperature control unit and permanent magnets. The outer conductor and the dielectric material of a RG 405/U type semi rigid coaxial cable are removed without scoring its inner conductor. This part is so called as *test cell*, where the MUT will be filled inside. The coaxial line is then placed into two split blocks to form an outer conductor for the test cell. The split blocks are made of brass material. They are plated with 2 μm gold electrode to reduce the conductor losses. The dimensions of the split blocks are given in Appendix A.4.1. The whole line can be treated as an artificial line, which includes three cascaded coaxial lines. These are a 70 mm long feeding line (RG 405/U), a 35 mm long test cell and again the 70 mm long feeding line. The inner conductor hangs on air without any mechanical support inside the test cell. This leads to a bending of the inner conductor, and its effects are discussed in the next section. Two filling channels are drilled into the split blocks to fill the MUT and to remove the air simultaneously.

The temperature of the setup is controlled by using Peltier modules, which is a thermoelectric cooler and heater [Sei+01]. The heat between the setup and the

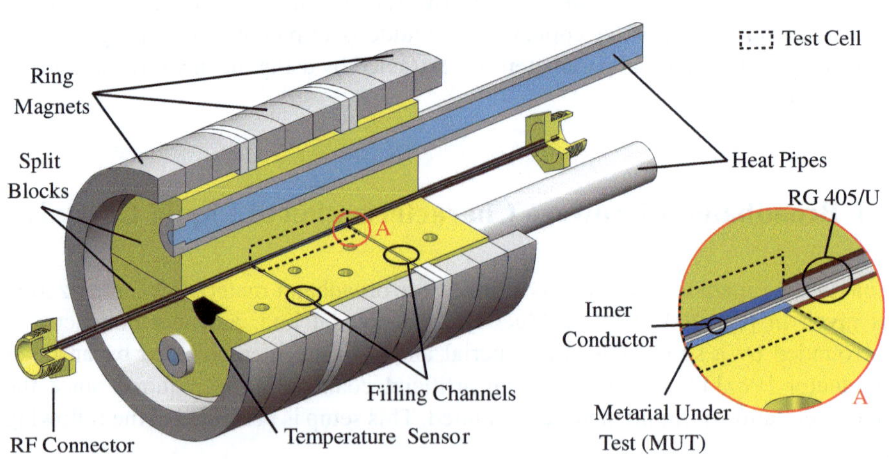

Fig. 3.1 Cut-view drawing of the temperature-controlled coaxial line setup

Peltier modules are transferred trough two heat pipes. An LM35H type temperature sensor is embedded into the split block. These sensors have a $\pm 0.5°$ uncertainty. They are located 22.5 mm far away from the test cell, resulting in additional thermal resistance. The total uncertainty in the temperature of the MUT is estimated to be less than $\pm 1.5°$. During the measurements, the current flow on the Peltier modules is controlled by ILX Lightwave temperature controller (LDT-5910).

The anisotropic LC molecules can be treated as isotropic and homogenous when the molecules are aligned in a certain direction. For this purpose, both electrostatic and magnetostatic fields are utilized herein. An electrostatic field is introduced by applying a bias voltage V_b across the inner and outer conductors. This aligns the molecules parallel to the electrical field of the coaxial line $\vec{E}$ and so the parallel complex relative permittivity of the LC material $\underline{\varepsilon}_{r,\mathrm{LC},\parallel}$ is effective. The magnetostatic field, on the other hand, is applied along the line to align the molecules perpendicular to the $\vec{E}$-field. As a result $\underline{\varepsilon}_{r,\mathrm{LC},\perp}$ is effective. The magnetostatic field is obtained by using permanent ring magnets surrounding the split block. The magnetic field is measured as 0.18T at position, where the test cell will be located.

3.1.2 Parameter Extraction

Extraction algorithms can be grouped into two as: *Point-by-point* and *multi-point* for this method [BJ+93]. Nicolson-Ross-Weir (NRW) [NR70, Wei74] method is a typical example, indeed forms a basis, for the point-by-point extraction techniques. This method works properly except for the frequencies, where the physical length of the line is equal to multiple of the half wavelength. At these frequencies, extracted results diverge because the phases of reflection coefficients can not be measured accurately, especially for low-loss materials [BJ+93]. The multi-point techniques avoid unphysical results since they allow correlations between the frequency measurements. They require optimization algorithms such as nonlinear least-square [BJ+93] or Newton-Raphson iteration [Mue07]. As is usual, an initial guest is essential to obtain correct values for optimization algorithms.

In this work, a point-by-point extraction method is utilized in two steps. In the first step, the complex relative permittivity $\underline{\varepsilon}_r$ is determined from the measured scattering parameters by using the Nicolson-Ross-Weir method, which is given in Appendix A.2.1. Then, these predetermined values are used as initial guesses to model the scattering parameters of the artificial coaxial as [BJ+93]

$$[S]_{Model} = \frac{1}{1-\underline{\Gamma}^2\underline{z}^2}\begin{pmatrix} \underline{\Gamma}\left(1-\underline{z}^2\right) & \underline{z}\left(1-\underline{\Gamma}^2\right) \\ \underline{z}\left(1-\underline{\Gamma}^2\right) & \underline{\Gamma}\left(1-\underline{z}^2\right) \end{pmatrix}. \tag{3.1}$$

with

$$\underline{\Gamma} = \frac{\sqrt{1/\underline{\varepsilon}_r} - 1}{\sqrt{1/\underline{\varepsilon}_r} + 1}, \tag{3.2}$$

$$\underline{z} = e^{-\underline{\gamma}\ell_{\text{physical}}} = e^{-(\omega \ell_{\text{physical}}/c_0)\underline{\varepsilon}_r}, \tag{3.3}$$

for each frequency. In Eq. (3.3), ℓ_{physical} is the physical length of the line, ω is the angular frequency and c_0 is the speed on the light in vacuum. The square error between the modelled and measured scattering parameters are minimized by optimizing the complex relative permittivity of the MUT. A nonlinear least-square method from MatLab is used for the optimization. The complex permittivity is decomposed into two real values by

$$\underline{\varepsilon}_r = \varepsilon_r(1 - j \tan\delta), \tag{3.4}$$

where ε_r is the relative permittivity and $\tan\delta$ is the loss tangent of the MUT.

The measured scattering parameters include two loss mechanisms. These are conductor losses caused by the finite conductivity of the electrodes and the dielectric losses because of the absorption of the radio frequency (RF) signal by the material. The conductor losses are taken into account as well while modelling the scattering parameters to perform an accurate extraction for the dielectric losses. The complex propagation constant of a coaxial line in Eq. (3.3) can be derived as

$$\underline{\gamma} = \ell_{\text{physical}}\sqrt{\omega\varepsilon_r\left((K\tan\delta - L) + j(L\tan\delta + K)\right)} \tag{3.5}$$

with

$$K = \frac{R_s\varepsilon_0(r_i + r_o)}{\ln(r_o/r_i)(r_i r_o)}, \tag{3.6}$$

$$L = \omega\mu_0\varepsilon_0 \tag{3.7}$$

by using the transmission line equivalent circuit (see Appendix A.3.1). The term K in Eq. (3.6) appears because of the finite conductivity of the electrodes. There, r_i and r_o are the inner and outer conductor radiuses, respectively. R_s is the surface resistivity with $R_s = \sqrt{\omega\mu_0/2\sigma}$, where σ is the conductivity of an electrode. In Eq. (3.7), μ_0 and ε_0 are the permeability and permittivity of free space, accordingly.

The pure conductor losses can be seen in $\underline{\gamma}$ by a product of ε_r and K (see Eq. (3.5)). The insertion losses caused only by the conductor losses not only depend on R_s (equivalently K), but also depend on the dielectric constant of the MUT since K is multiplied by ε_r. Therefore, the conductor losses have to be considered in each iteration of the optimization procedure. To do that, R_s has to be known to calculate K (see Eq. (3.6)). R_s can be determined from the characteristic impedance of the empty test cell, namely $Z_{\text{Line, Air}}$. Based on the measured scattering parameters of the empty test cell, $Z_{\text{Line, Air}}$ can be written as [EE92]

$$Z_{\text{Line, Air}} = 50\sqrt{\frac{(1 + S_{11})^2 - S_{21}^2}{(1 - S_{11})^2 - S_{21}^2}}. \tag{3.8}$$

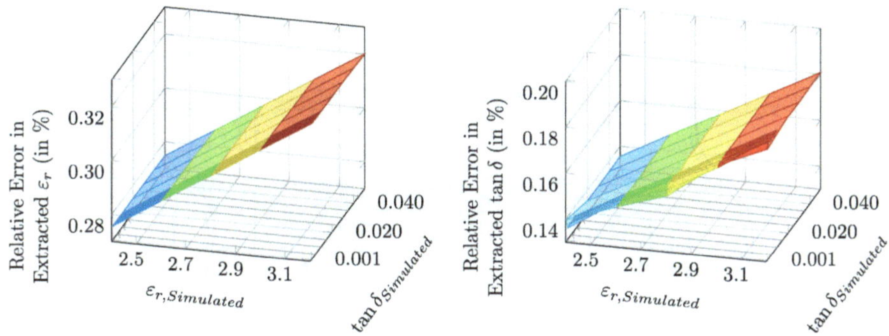

Fig. 3.2 *Left:* Relative error between the extracted and simulated relative permittivities ε_r. *Right:* Loss tangents $\tan\delta$ (*right*), both at 20 GHz. To calculate the relative errors, the difference between two values are normalized with respect to the simulated one

On the other hand, $Z_{\text{Line, Air}}$ can also be calculated as

$$Z_{\text{Line, Air}} = \frac{\sqrt{L - jK}}{2\pi\varepsilon_0\sqrt{\omega}} \ln(r_0/r_i), \tag{3.9}$$

by using the transmission line equivalent circuit model given in Appendix A.3.1 and by assuming that $\underline{\varepsilon}_{r,\text{Air}} = 1$. Since Eqs. (3.8) and (3.9) are equal to each other, the unknown K can be determined for each frequency. Thus, the complex propagation constant of the coaxial line in Eq. (3.5) can be calculated for a given $\underline{\varepsilon}_r$ considering the conductor losses. Then, it is used in Eq. (3.1) to model the scattering parameters of the test cell accurately.

The accuracy of the proposed algorithm is confirmed by extracting $\underline{\varepsilon}_r$ form simulated scattering parameters. For this purpose, the coaxial line is simulated in CST Microwave Studio for typical values of ε_r and $\tan\delta$ of a microwave LC material. The conductor losses are considered as well by assigning the electrodes with lossy copper material ($\sigma_{\text{copper}} = 5.8 \times 10^7$ S/m). In Fig. 3.2, relative error between the extracted and simulated values are given at 20 GHz. This is a systematic error increasing with the relative permittivity since the electrical length of the line increases as well. It should be noted that there are two sources for these errors. The first one is the accuracy of the proposed method and the second one is the accuracy of the CST simulations. Nevertheless, no further investigations are done to distinguish the two error sources since the overall errors are less than 0.3 % for the both ε_r and $\tan\delta$.

3.1.3 *Effects of Inner Conductor Bending*

There is no mechanical support to hold the inner conductor straight and in the center of the test cell. This leads to a bending, whose effects on the extracted parameters

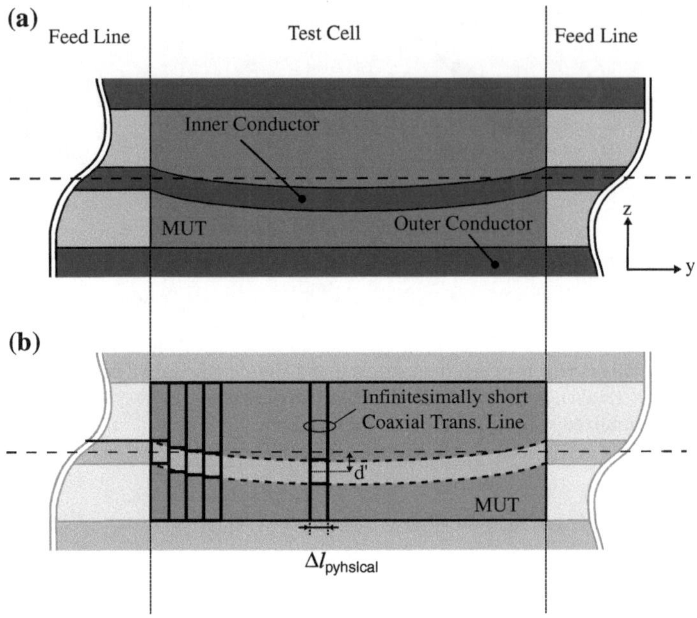

Fig. 3.3 **a** Side view of an coaxial line, whose inner conductor is bended in a banana-shaped, **b** which is modelled with infinitesimally short eccentric transmission lines

are discussed herein. For the analysis, it is assumed that the inner conductor bends in a banana-shaped as shown in Fig. 3.3a. However, no bending occurs in the feeding line sections since there, the inner conductor is surrounded by the solid dielectric.

As it is depicted in Fig. 3.3b, such a transmission line can be described as serial combination of infinitesimally short coaxial transmission lines. These lines have certain amount of offsets d' in their inner conductors so that, they are not concentric structures like a conventional coaxial line. These lines are so called as *eccentric coaxial lines*. The maximum offset, which is denoted as d'_{max}, stays in the middle of the test cell. The scattering parameters of the banana-shaped transmission line can be determined by cascading the transmission matrixes (ABCD matrixes) of the several eccentric coaxial lines. The eccentric coaxial lines are assigned to be infinitesimally short with a physical length of $\Delta\ell_{\mathrm{physical}}$ to enhance the modelling accuracy.

Electrical properties of the eccentric coaxial lines can be calculated by using either numerical formulas [YA66] or analytical formulas [MS88, Kra99, Ser06], which transforms the given geometry into another one. In this work, the conformal mapping technique is utilized. As illustrated in Fig. 3.4, the eccentric coaxial lines are mapped onto the concentric ones by using [Ser06]

$$w = \frac{z - a}{za - 1}. \tag{3.10}$$

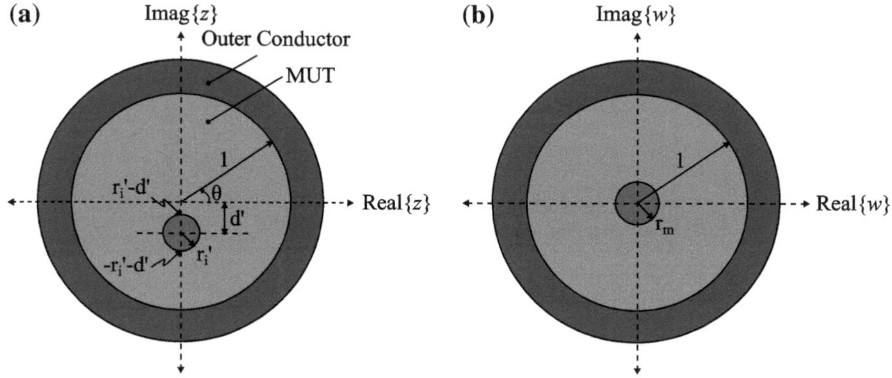

Fig. 3.4 Cross-sectional view of **a** an eccentric coaxial line is transformed into **b** a concentric coaxial line by using the conformal mapping technique. After the mapping, the inner conductor radius is equal to r_m

In Eq. (3.10), z is in the physical domain defined by the eccentric coaxial line and w is in the canonical domain, where the concentric line is obtained. It should be noted that Eq. (3.10) enables to preserve the outer conductor still in the center of the coordinate systems while transforming the inner conductor to the center. This is accomplished by normalizing the line dimensions, e.g. the inner conductor radius r_i, with respect to the outer conductor radius r_o. From the electrical point of view, the normalization is reasonable when the metallic losses are neglected. Given that, the characteristic impedance and the propagation constant of a coaxial line can be calculated by using ratio r_i/r_o instead of using absolute values of r_i and r_o. This can be seen from the formulas giving the lumped elements used in the equivalent circuit [Poz11] (and also refer Appendix A.3.1). These elements depend on r_i/r_o except for the serial resistance, representing the metallic losses.

The normalized outer conductor becomes a unit circle, which can be expressed by $z = e^{j\theta}$ with θ is from 0° to 360°. When the mapping is performed, it is still a unit circle with $w = e^{j\theta+\theta_a}$ for any value of a. Here, θ_a is a rotation angle depending on the chosen real-valued constant a. Indeed, the rotation has no impact on the electrical properties of the coaxial line because of the shape symmetry. The normalized inner conductor radius is denoted as r_i'. The absolute value of the inner conductor offset d' depends on the position of the chosen eccentric line (see Fig. 3.3). As aforementioned, the physical length of the test cell ℓ_{physical} is divided into N number of eccentric coaxial lines, which are then $\Delta\ell_{\text{physical}}$ long. In this case, the banana-shaped bending can be modelled by using a cosine function such as

$$d' = d'_{\text{max}} \cos\left[\pi\left(\frac{n}{N} - \frac{1}{2}\right)\right], \tag{3.11}$$

where n is an integer from 0 to N and $\Delta\ell_{\text{physical}} = \ell_{\text{physical}}/N$. After determining d', for instance, the points $r_i' - d'$ and $-r_i' - d'$ (see Fig. 3.4) are transformed into

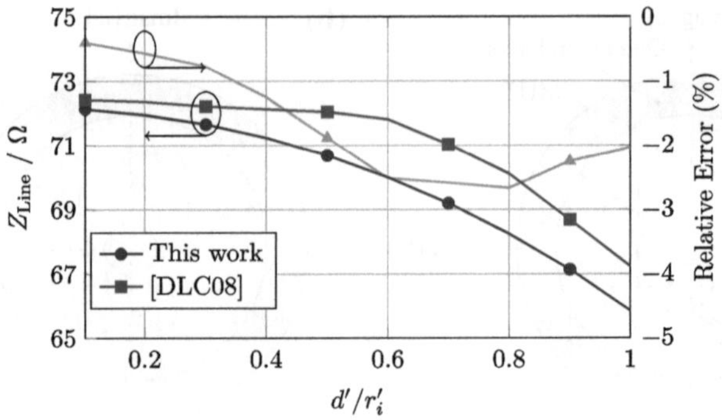

Fig. 3.5 Characteristic impedance of the empty test cell ($\varepsilon_r = 1$) versus the relative inner conductor offset d'/r_i'. The inner and the outer conductor radiuses are $r_i = 0.255$ mm and $r_o = 0.850$ mm, respectively. The relative error is calculated by normalizing the difference between the two curves with respect to the reference value for each d'/r_i'

$-r_m$ and r_m, respectively. r_m is the radius of the inner conductor in the canonical domain. This leads to two possible solutions of a. The right one is chosen in such a way that $|w| < 1$ is provided because of the normalization.

To verify the accuracy of the presented method, the characteristic impedance Z_{Line} of the empty test cell is calculated by using Eq. (3.10) for different offset values. These results are compared to the ones given in [DLC08], which are determined by using an numerical boundary element method. The characteristic impedances and the relative error in between are given in Fig. 3.5. Both results match quite well with a maximum error of less than 2.7 %.

Based on this method, the inner conductor radius of the mapped coaxial line r_m is calculated as

$$r_m = \frac{u^2 + \sqrt{(1-uv)^2 - 4(d')^2} - 1}{u(1-uv) - u\sqrt{(1-uv)^2 - 4(d')^2} - 2d'} \tag{3.12}$$

with

$$u = r_i' + d' \qquad \text{and} \qquad v = r_i' - d'.$$

For the investigations, a MatLab script is written, which models the banana-shaped bending for a given d_{max} in the first step. Then, the inner conductor offsets are calculated for N number of eccentric coaxial lines. Each line is mapped onto the concentric one to calculate the resultant line impedance and the propagation constant. In the next step, N number of transmission matrixes are calculated and cascaded to obtain that of the bended line. These parameters are converted into the scattering parameters [Fri94], which are then used to extract the material parameters as described in Sect. 3.1.2.

It is assumed that the test cell is filled with a material having $\varepsilon_r = 3$ and $\tan\delta = 0.01$. Its scattering parameters are determined for different maximum offsets

$d'_{\max}$. Figure 3.6 shows the relative error between the extracted values and the original values. The results show that the bending of the inner conductor hardly affects the extracted parameters. There are two reasons for these results. First, the characteristic impedance and the propagation constants of adjacent eccentric lines are close to each other because of the banana-shaped bending. Hence, no discontinuity occurs through the test cell. Second, $d'_{\max}/r_i$ is small, which is estimated to be 1.5 by measuring the dimensions under a microscope. Additionally, it can be seen that the higher the $d'_{\max}/r_i$ ratio the lower the extracted ε_r since the error in ε_r is lower than 0 and its absolute value increases. This means that the line becomes electrically smaller as the inner conductor bends.

3.1.4 Measurement Setup and Results

The complete setup is shown in Fig. 3.7. The semi rigid cable operates up to 64 GHz in the fundamental TEM mode according to formulas given in [Poz11]. This frequency is defined by the line geometry and its dielectric material. Nevertheless, the operating frequency of the whole setup is limited up to be 45 GHz because the first higher order mode appears, for instance, at 48 GHz when the test cell is filled with an LC sample with $\varepsilon_{r,\mathrm{LC}} = 3.5$. The feeding lines are installed with K-connectors from Anritsu, which operates properly up to 45 GHz as well. Effects caused by the connectors and the feed lines are eliminated by using a trough-reflect-line (TRL) calibration technique [CS07], enabling to transform the calibration reference planes to the boarders of the test cell. This calibration technique requires at least two line standards to perform the measurements from dc to 45 GHz. Nonetheless, only one calibration standard is realized, which enables to perform the measurements from 3.5 to 30 GHz.

The LC materials measured in this work are supplied by company Merck KGaA, Darmstadt. The extracted material parameters of two LCs, namely $GT3-23001$

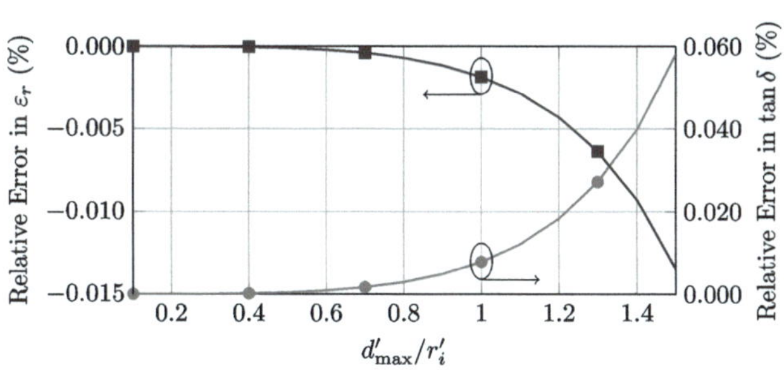

Fig. 3.6 Relative error in extracted **a** relative permittivity ε_r and **b** loss tangent $\tan\delta$. The reference values are $\varepsilon_r = 3$ and $\tan\delta = 0.01$. The relative errors are calculated by normalizing the difference between the extracted and the reference values with respect to the reference values. The test cell is divided into $N = 50$ sections through the line

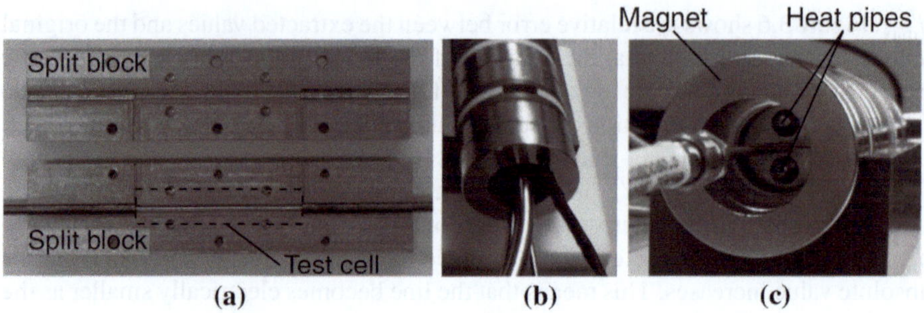

Fig. 3.7 Photographs of the realized artificial coaxial line setup: **a** dissembled split blocks, **b** heat pipes and the permanent magnets, **c** split block located inside the magnets

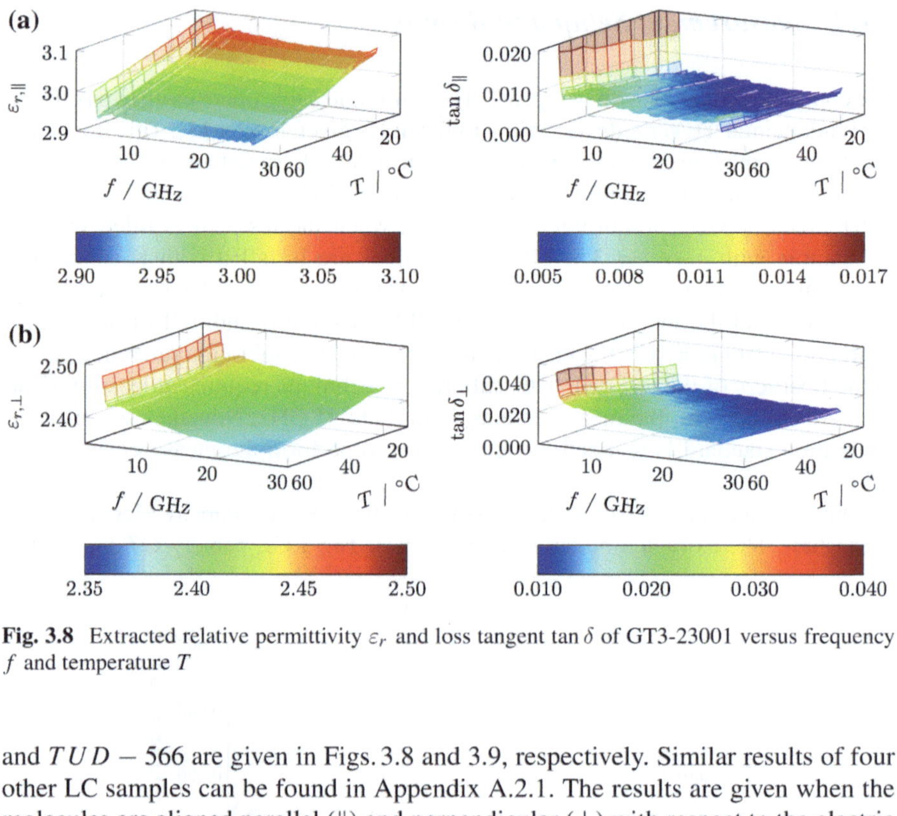

Fig. 3.8 Extracted relative permittivity ε_r and loss tangent $\tan\delta$ of GT3-23001 versus frequency f and temperature T

and $TUD-566$ are given in Figs. 3.8 and 3.9, respectively. Similar results of four other LC samples can be found in Appendix A.2.1. The results are given when the molecules are aligned parallel ($\|$) and perpendicular ($\perp$) with respect to the electric field of the coaxial line. After aligning the molecules, the scattering parameters are measured while heating the setup from 10 to 60 °C in 5 °C steps.

The material parameters are almost flat or changes slightly over the frequency as well as over the temperature. In particular, no phase transition is detected from 10 to 60 °C. These results clearly show the temperature range for the nematic phases are extended for the microwave LCs compared to the results published in [Mue07].

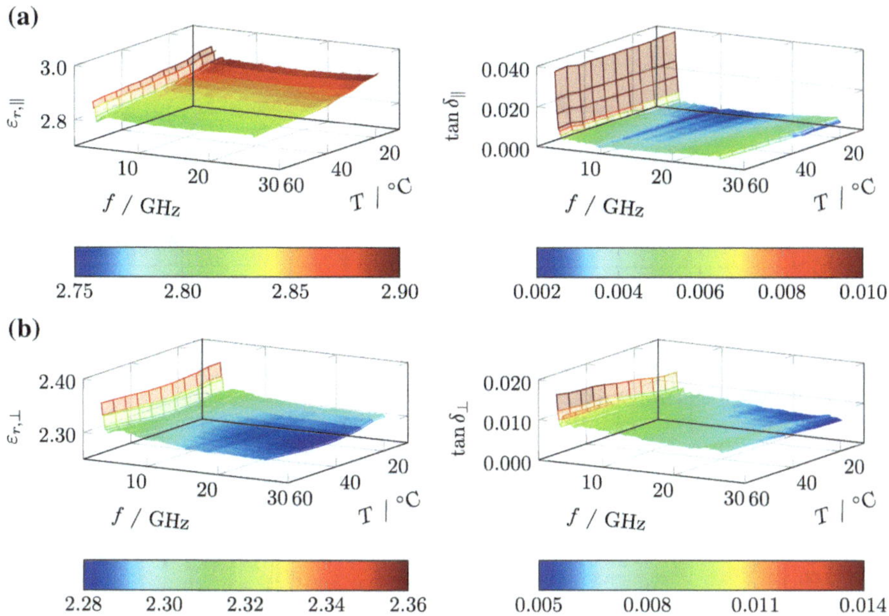

Fig. 3.9 Extracted relative permittivity ε_r and loss tangent $\tan\delta$ of GT3-23001 versus frequency f and temperature T

This is accomplished by doping functional materials into host LC materials [Goe09]. Figure 3.10 shows two essential material parameters, which are the material tunability $\tau_{\mathrm{LC}} = (\varepsilon_{r,\mathrm{LC},\parallel} - \varepsilon_{r,\mathrm{LC},\perp})/\varepsilon_{r,\mathrm{LC},\parallel}$ and the material efficiency $\eta_{\mathrm{LC}} = \tau_{\mathrm{LC}}/\tan\delta_{\perp}$ (see Sect. 2.1.1). Both GT3-23001 and TUD-566 type LCs have almost same amount of tunability. This is about 0.2 and increases only slightly with increasing frequency. On the other hand, the efficiency of the TUD-566 LC sample is approximately twice that of the GT3-23001 sample. This is because of the lower dielectric losses of the TUD-566 sample. Additionally, both efficiencies increase with increasing frequency. This makes the LC materials attractive for microwave or even higher frequencies since their dielectric losses decrease at higher frequencies unlike to other technologies.

The temperature dependencies of τ_{LC} and η_{LC} are given in Fig. 3.10 as well. Both materials are always in the nematic phase in the temperature of interest since $\tau_{\mathrm{LC}} > 0$. But these tunabilities reduce as the temperature rises. For instance, the tunability of the GT3-23001 sample reduces from 0.204 to 0.185 if the temperature is increased from 10 to 60 °C. Similarly, the material efficiencies decrease over the temperature. However, η_{VDL} decreases faster than τ_{LC}, implying that the dielectric losses are more sensitive to the temperature change compared to the relative permittivities.

Beside the material characterization, an LC filled coaxial line can be used as a variable delay line. Indeed, the delay lines are subject to the following chapters. Nevertheless, this is discussed here briefly. For this purpose, the differential phase shift $\Delta\phi_{\beta}$ and the delay line efficiency η_{VDL} are plotted in Fig. 3.11

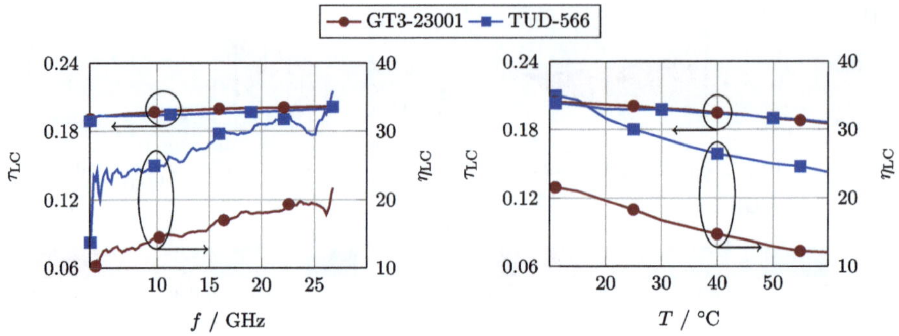

Fig. 3.10 *Left* Extracted liquid crystal material tunability τ_{LC} and efficiency η_{LC} versus frequency at 25 °C. *Right* τ_{LC} and η_{LC} versus temperature at 20 GHz

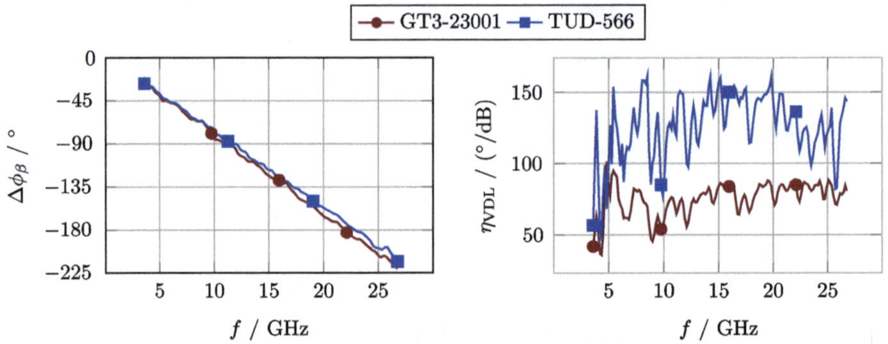

Fig. 3.11 *Left* Measured differential phase shift $\Delta\phi_\beta$ of LC filled coaxial line, used as a variable delay line. *Right* The resultant delay line efficiencies η_{VDL}. Both results are given at 25 °C

at 25 °C. For instance, if the test cell is filled with any of the two LC samples a $\Delta\phi_{\beta,\max}$ of 155° can be provided at 20 GHz since the both samples have almost the same amount of tunability. The delay line efficiencies are again different because of different material efficiencies. The relationship between the two efficiencies can be approximated by $\eta_{VDL} \approx 6.6 \times \eta_{LC}$ [Mue07]. This leads to theoretical delay line efficiencies of 188.8 and 204.6 °/dB at 20 GHz if the test cell is filled with GT3-23001 and TUD-566 LC samples, respectively. On the other hand, the measured values are 70.86 and 133.2 °/dB, accordingly, at the same frequency. The measured ones are less than the theoretical ones because of the additional metallic losses. Given that, these values are higher compared to the planar delay lines when the same LC sample is used (see Sect. 4.2). This is due to the fact that the effective tunability experienced by the RF field is equal to the available material tunability in a coaxial line since the line is shielded. However, for the planar delay lines, there is always fringing fields, which means that a part of the RF field propagates through a not tunable material. The drawback of the coaxial delay line is the magnetic alignment of the molecules. This has to be substituted with an electrical alignment method to be a practical solution.

3.2 Microwave Characterization of Glasses

This section presents characterization of glass substrates for microwave frequencies. Two methods are utilized to extract the relative permittivity ε_r and the loss tangent $\tan\delta$ from the measured scattering parameters. In a first step, a *split-cylinder-cavity* [Che+04] is used.[1] This method requires no sample preparation and well suited for planar substrates. Therefore, it is used for the first and quick measurements to figure out promising low-loss glass substrates. Its drawback is that the measurements are done only at one frequency that is around 20 GHz in this case. In a second setup, the promising glass substrates are characterized once again over a broad frequency range. For this purpose, a coplanar waveguide is located on a glass substrate and its scattering parameters are measured from dc to 30 GHz.

3.2.1 Split-cylinder-cavity Method

The split-cylinder-cavity method was first proposed by Kent [Ken96] to characterize sheet dielectric substrates. A cylindrical cavity is split into two parts and the material under test is placed between the halves [Che+04]. The cavity is designed to operate at the TE_{011} mode and therefore, the electrical field strength is maximized in the middle of the cavity that is on the MUT simultaneously. Figure 3.12 shows the transmission coefficients of the empty cavity. The cavity is excited by using a magnetic coupling method [Che+04], where a inner conductor of a coaxial cable is soldered onto the cavity's inner wall. The desired TE_{011} mode resonates at $f_r = I22.803$ GHz depending on dimensions of the cavity. The quality factor is calculated as 700. Although being sufficient for the characterization, the quality factor is quite less since the effects of the magnetic coupling and SMA connectors installed to the coaxial cable cannot be extracted by the calibration.

If the cavity is loaded with the substrate, the material parameters can be extracted from the resultant resonance frequency and quality factor. The theory of the parameter extraction is summarized in Appendix A.2.2. In the frame of this work, six glass samples are characterized at the room temperature. These results are listed in Table 3.1.

The fused silica is a typical glass substrate used in microwave applications. It has low dielectric losses whereas its size is limited to 10×10 cm by manufactures. Therefore, it is not applicable for large antenna arrays. On the other hand, AF32, AF45 and BF33 glasses have still moderate dielectric losses (<0.01). It should be

[1] The setup is not realized by the author.

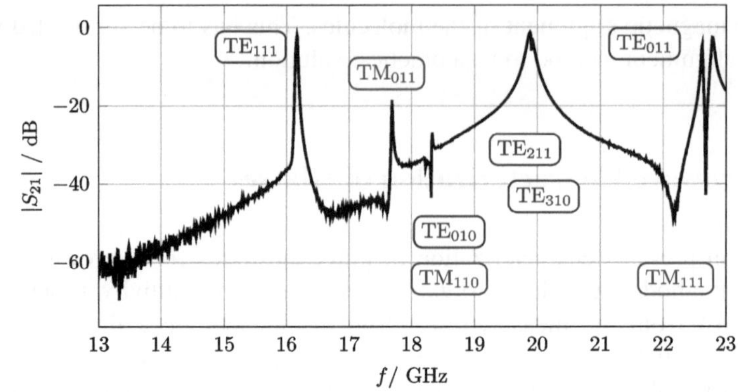

Fig. 3.12 Magnitudes of the measured transmission coefficients of an empty split-cylinder-cavity versus frequency. The diameter and the height of the cavity are 20 and 11 mm, respectively

mentioned that AF45 and BF33 resonate at the same frequency even though they have different relative permittivities. This is because of their different thicknesses, resulting in the resonance at the same frequency. These three glasses are considerably cost-effective compared to the fused silica and they can be provided in larger sizes. Both D263 T and B 270 type glass have high dielectric losses and so that, they are not convenient for an application. These results imply that D263 T and B 270 feature dispersions when the results in Table 3.1 are compared. This could be critical in an application, since the frequency sensitivity of a device is enhanced.

Based on these results, BF33, AF32 and AF45 type glasses are chosen for a further characterization to figure out their dispersion characteristics. These results are presented in the following part.

Table 3.1 Extracted material parameters by using the split-cylinder-cavity method

Sample	Thickness (mm)	Data from manufacturers at 1 MHz		Extracted parameters at f_r/GHz		
		ε_r	$\tan\delta$	f_r	ε_r	$\tan\delta$
Fused silica [SCHd]	0.30	3.8	0.0014	20.98	3.784	0.0015
AF32 [SCHa]	0.70	5.1	-	18.825	5.227	0.0086
AF45 [Af4]	0.51	6.2	0.0009	18.210	5.868	0.0083
BF33 [SCHc]	0.70	4.6	0.0037	18.210	4.529	0.0080
D263 T [D26]	0.55	6.7	0.0061	18.65	5.017	0.0313
B 270 [SCHb]	0.52	7.0	-	17.27	5.361	0.0242

3.2.2 On-wafer Characterization Method

The broadband material parameters of AF32, AF45 and BF33 type glasses are extracted from the scattering parameters of a coplanar waveguide (CPW), which is implemented on these glasses. A photograph of one of the realized structures is shown in Fig. 3.13. The reason for choosing the CPW line is that it allows using an on-wafer measurement facilities without any additional transition. Its electrodes are realized by using lithography processes (see Sect. 2.2.1). The characteristic impedance of the line is equal to 50 Ω. The scattering parameters are measured from 0.4 to 30 GHz. The magnitudes of the reflection and transmission coefficients are plotted in Fig. 3.14.

The CPW lines are matched quite well at the whole frequency of interest with a maximum insertion of less than 1 dB. These results already indicate a low dispersion characteristic for each glass substrate since no abrupt reflection or resonance is detected. Nevertheless, the material parameters can be determined quantitatively by using the formulas presented in Appendix A.3.2. Extracted relative permittivity ε_r and loss tangent $\tan\delta$ are given in Fig. 3.15.

These results match quite well to the previous ones. Particularly, both results are compared in Table 3.2 at 18.5 GHz. The relative differences are calculated by normalizing the difference between the two results with respect to the corresponding result obtained by using the split-cylinder-cavity method. ε_r values determined by using the CPW line is lower than those obtained by using the split-cylinder-cavity method. In the CPW line method, first the effective relative permittivity is determined from the measured scattering parameters. These values are lower than the actual relative permittivities of the substrates since the CPW line is surrounded by air. Later on, the actual ε_r are calculated from the effective ones by considering the CPW geometry. Any fabrication tolerances cause uncertainty in the final results. Nevertheless, the relative differences are less than 3 %.

In contrary, the loss tangents obtained by using the CPW line are higher than the previous values. This could originate because of two reasons. First, the resolution of

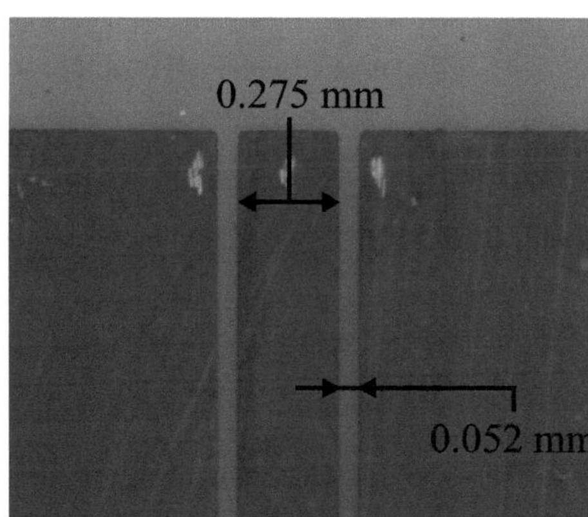

Fig. 3.13 A photograph of the coplanar waveguide realized on BF33. The length of the line is 10 mm and the height of the substrate is 0.7 mm

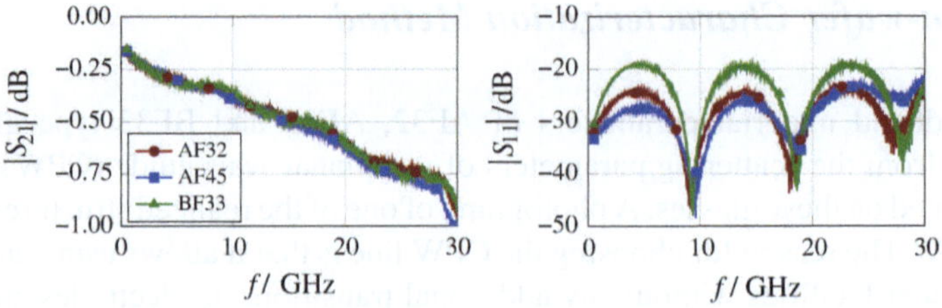

Fig. 3.14 *Left:* Magnitudes of measured transmission of the coplanar waveguides. *Right:* Magnitudes of measured reflection coefficients

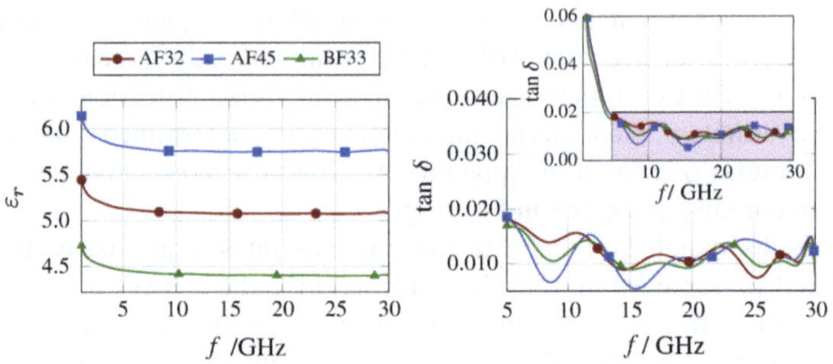

Fig. 3.15 *Left:* Extracted relative permittivity ε_r f glass substrates. *Right:* Extracted loss tangents $\tan\delta$

Table 3.2 Extracted material parameters by using the coplanar waveguide

Sample	Extracted parameters at 18.5 GHz		Relative difference compared to Table 3.1	
	ε_r	$\tan\delta$	in ε_r (%)	in $\tan\delta$ (%)
AF32 [SCHa]	5.079	0.0108	2.83	−25.6
AF45 [Af4]	5.752	0.0104	1.97	−25.3
BF33 [SCHc]	4.402	0.0093	2.80	−15.85

the split-cylinder-cavity method is higher since being a resonance method. Therefore, the dielectric losses are extracted more accurately. Second, the CPW lines have high metallic losses. Indeed, they are extracted while calculating the material parameters by using analytical formulas (see Appendix A.3.2). However, these formulas are approximations depending on the device geometry.

3.3 Summary

This chapter presented the microwave characterization of liquid crystal materials and glass substrates. Three characterization methods were studied to extract the relative permittivity ε_r and the loss tangent $\tan\delta$. These methods are the coaxial line method used for liquid crystal materials, the split-cylinder-cavity method and the coplanar waveguide based on-wafer characterization method used for glass substrates.

Several LC samples were characterized from 3.5 to 30 GHz while controlling the temperature of the setup between 10 and 60 °C. Table 3.3 lists the extracted material parameters at 30 GHz and at the room temperature. Similar results are given in the same table, which are obtained by using the cavity perturbation method [Goe09].

Both methods have their own systematic errors as discussed in [Che+04]. This leads to, for instance, extract lower ε_r and $\tan\delta$ when the coaxial line method is utilized. As a result, the relative difference between the material tunabilities τ_{LC} are about 20 %. For the material efficiencies η_{LC}, on the other hand, the results cannot be compared that easily because of two reasons. First, the loss tangents are much more sensitive to any uncertainty and tolerances than ε_r. Second, although providing more information over a frequency band, the broadband systems are less accurate for low-loss materials. This is due to the fact that the uncertainties in the measured scattering parameters increase in case of a low-loss material [BJ+93].

Beside the LC materials, several glass substrates were characterized to determine low-loss and cost-effective substrates. For this purpose, glass substrates were characterized by using the split-cylinder-cavity method in the first step. The ones having $\tan\delta \leq 0.01$ were further characterized in the second step. A coplanar waveguide was realized on the substrates for the broadband characterization. As a results, three glasses were chosen for possible device fabrication. These substrates were AF32, AF45 and BF33. In particularly, BF33 features 10 times higher dielectric losses than a fused silica. In fact, its dielectric losses are not so high and actually they are comparable with those of the LC materials. Therefore, device performances are not degraded significantly since the electromagnetic signals propagate mainly through

Table 3.3 Extracted material parameters at the room temperature. The results given in bold are extracted at 19 GHz and all the others are extracted at 30 GHz

LC material	Alignment	Cavity perturbation [Goe09]				This work			
		ε_r	$\tan\delta$	τ_{LC}	η_{LC}	ε_r	$\tan\delta$	τ_{LC}	η_{LC}
K15	$\parallel$	2.91	00132	0.113	3.99	2.80	0.0076	0.090	6.08
	$\perp$	2.58	0.0294			2.55	0.0148		
TUD-862	$\parallel$	2.66	0.0041	0.128	9.98	2.59	0.0040	0.102	16.26
	$\perp$	2.32	0.0128			2.32	0.0063		
TUD-566	$\parallel$	3.11	0.0021	0.254	38.49	2.86	0.0028	0.202	35.9
	$\perp$	2.32	0.0066			2.28	0.0056		
GT3-23001	$\parallel$	**3.28**	**0.0038**	**0.252**	**17.48**	3.02	0.0053	0.205	21.72
	$\perp$	**2.46**	**0.0143**			2.40	0.0094		
TUD-670	$\parallel$	**3.06**	**0.0037**	**0.210**	**20.11**	2.83	0.0027	0.167	23.43
	$\perp$	**2.42**	**0.0104**			2.36	0.0071		
TUD-899	$\parallel$	**3.41**	**0.0025**	**0.279**	**40.37**	3.06	0.0042	0.215	34.01
	$\perp$	**2.46**	**0.0069**			2.40	0.0063		

the LC layer. On the other hand, BF33 enables mass productions because of its low price and large available sizes both compared to the fused silica.

References

[Af4] AF45 alkali free thin glass. Praezisions Glas and Optik GmbH, http://www.pgo-online.com/intl/katalog/AF45.html

[BJ+93] J. Baker-Jarvis, M.D. Janezic, J.H. Grosvenor, R.G. Geyer, Transmission/reflection and short-circuit line methods for measuring permittivity and permeability. Technical report, National Institute of Standards and Technology, 1355-R, 1993

[Che+04] L.F. Chen, C.K. Ong, C.P. Neo, V.V. Varadan, V.K. Varadan, *Microwave Electronics: Measurement and Materials Characterization* (Wiley, New York, 2004)

[CS07] R.J. Collier, A.D. Skinner (eds.), *Microwave Measurements*, 3rd edn. (IET Electrical Measurement Series) (The Institution of Engineering and Technology, London, 2007)

[DLC08] X. Dong, L. Mai, X. Chen, Computation of the characteristic impedance of eccentric coaxial transmission lines by boundary element method, in *Proceedings of World Automation Congress WAC, 2008*, pp. 1–5, 2008

[D26] D263 T borosilicate thin glass. Praezisions Glas and Optik GmbH, http://www.pgo-online.com/intl/katalog/D263.html

[EE92] W.R. Eisenstadt, Y. Eo, S-parameter-based ic interconnect transmission line characterization. IEEE Trans. Compon. Hybrids Manuf. Technol. **15**(4), 483–490 (1992)

[Fri94] D.A. Frickey, Conversions between s, z, y, h, abcd, and t parameters which are valid for complex source and load impedances. IEEE Trans. Microw. Theory Tech. **42**, 205–211 (1994)

[Goe09] F. Goelden, Liquid crystal based microwave components with fast response times: material, technology, power handling capability. PhD thesis, Technische Universitaet Darmstadt, 2009

[Ken96] G. Kent, Nondestructive permittivity measurement of substrates. IEEE Trans. Instrum. meas. **45**, 102–106 (1996)

[Kra99] G.S. Krantz, *Handbook of Complex Variables* (Birkhauser, Boston, 1999)

[MS88] P. Moon, D.E. Spencer, *Field Theory Handbook: Including Coordinate Systems, Differential Equations and Their Solutions* (Springer, New York, 1988)

[Mue07] S. Mueller, Grundlegende Untersuchungen steuerbarer passiver Flüssigkristall-Komponenten für die Mikrowellentechnik. PhD thesis, Technische Universitaet Darmstadt, 2007

[NR70] A.M. Nicolson, G.F. Ross, Measurement of the intrinsic properties of materials by time-domain techniques. IEEE Trans. Instrum. Meas. **19**(4), 377–382 (1970)

[Poz11] D.M. Pozar, *Microwave Engineering*, 4th edn. (Wiley, New York, 2011)

[SCHa] SCHOTT. AF 32 thin glass, http://www.schott.com/special_applications/english/download/schott_af32_eco_thinglass_july_2011_data_sheet_eng.pdf?PHPSESSID=d653rl07j4vqd3hpi2l5b5f0t6

[SCHb] SCHOTT. B 270 flat glass, http://www.schott.com/special_applications/english/download/schott_b_270_flat_glass_june_2012_eng.pdf

[SCHc] SCHOTT. Schott borofloat 33, http://www.schott.com/borofloat/english/download/borofloat_33_e.pdf

[SCHd] SCHOTT. Synthetic fused silica, http://www.schott.com/korea/korean/download/datasheet_fused_silica_.pdf

[Sei+01] W. Seifert, M. Ueltzen, C. Strumpel, W. Heiliger, E. Muller, One-dimensional modeling of a peltier element, in *Proceedings of XX International Conference on Thermoelectrics ICT 2001*, pp. 439–443, 2001

[Ser06] B.E. Sernelius, Electrodynamics, TFYY67, NFYD70. Linkoping Institute of Technology Department of Physics, Chemistry and Biology, http://www.ifm.liu.se/courses/TFYY67/tenta060317.pdf (2006)

[Wei74] W.B. Weir, Automatic measurement of complex dielectric constant and permeability at microwave frequencies. Proc. IEEE **62**(1), 33–36 (1974)

[YA66] H.Y. Yee, N.F. Audeh, Cutoff frequencies of eccentric waveguides. IEEE Trans. Microw. TheoryTech. **14**(10), 487–493 (1966)

Chapter 4
2D Electronic Beam Steering Phased Array

Phased arrays are one of the well-known types of beam steering antennas. Main beam direction of a phased array can be steered by means of mechanical or electrical devices as well as by means of a combination of both methods. Electronic beam steering phased arrays are in the scope of this work since they are robust, compact and fast. Such antennas consist of a distribution network, electronically tunable variable delay lines or phase shifters, receive/transmit modules (for active arrays) and radiating elements. The phase of each radiating element is adjusted to predefined phase values to tilt the radiated phase front in a specified direction [Bal05].

This chapter presents a two-dimensional (2D) electronic beam steering phased array in liquid crystal technology. Variable delay lines or phase shifters play an essential role concerning the performance, cost, and dimensions of such an antenna. Therefore, several planar LC delay lines are studied, realized and measured before a complete phased array is presented. In the first section, these delay lines are classified into two main topologies, namely transmission line and loaded line topologies. Underlying design parameters of different topologies are discussed. In the second section, four planar delay lines are designed and their measurement results are presented. Since low profile antennas are aimed for, the delay lines are realized by utilizing planar transmission lines such as coplanar waveguide, slotline and microstrip line, although they have higher losses compared to a hollow waveguide. Special attention has been given to improve device topologies so that the material tunability can be utilized with maximum efficiency and with lowest insertion loss. At the end, these devices are compared with respect to their insertion loss, compactness, response time and reliability. Particularly, it is stated that in which manner these parameters affect the antenna performance. In the third section, an antenna concept is proposed for the 2D electronic beam steering phased array in LC technology. Its realization and measurement results are presented. The proposed antenna advances the state-of-the-art for the LC based steerable antennas, which are limited to scan only in 1D [SHK09, HU+08]. Finally, significant results obtained in this chapter are summarized in the fourth section.

O. H. Karabey, *Electronic Beam Steering and Polarization Agile Planar Antennas in Liquid Crystal Technology*, Springer Theses, DOI: 10.1007/978-3-319-01424-1_4,

4.1 Liquid Crystal Based Variable Delay Line Topologies

There are two generic types of phase shifters: True phase shifter and true delay line [RL01]. The first type provides a flat differential phase shift $\Delta\phi_\beta$ within its defined bandwidth. However, $\Delta\phi_\beta$ of the second type is linearly proportional to an operating frequency f_0 and therefore, it is convenient for broadband systems. Variable delay lines are preferred, specifically, for the phased arrays to prevent *beam squinting*, which is defined as the deviation of the beam direction caused by the variations of the operating frequency [RL01, Bal05, Gol01, Bha06]. On the other hand, an LC device can be utilized as either a true phase shifter or a true delay line. Figure 4.1 depicts the typical frequency dependent differential phase shift characteristic of the LC device, i.e. inverted microstrip line (IMSL). The frequency dependency of the device can be compensated by tuning the applied bias voltage V_b that controls the relative permittivity of the LC layer $\varepsilon_{r,\mathrm{LC}}$ continuously. For instance, in a given bandwidth (see ❶ in Fig. 4.1) the bias voltage is tuned from $V_{b,i}$ to $V_{b,j}$ to obtain a true phase shifter characteristic. All LC devices discussed in this work are intentionally named as variable delay lines (VDL) since phased arrays are of interest.

LC based variable delay lines can be classified into two topologies. These are namely *transmission line* and *periodically loaded line* topologies. The transmission line topology refers to a transmission line, whose dielectric is completely made of a tunable LC material like in IMSL topology. The periodically loaded line approach assumes that a delay line consists of periodic structures that are composed of pieces

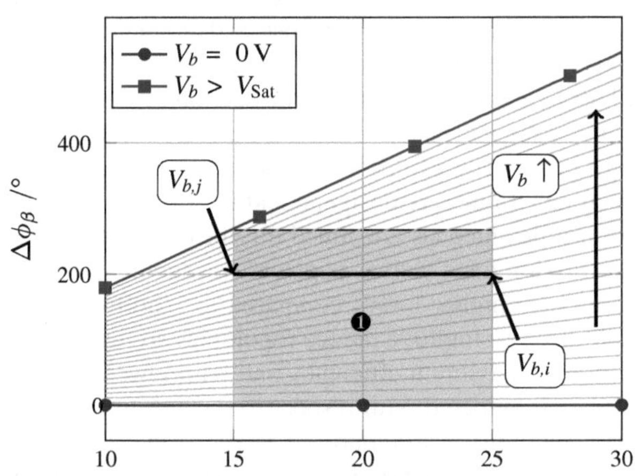

Fig. 4.1 Typical differential phase shift versus frequency characteristic of a LC phase shifter. The device operates as a true delay line inherently. To operate as a true phase shifter at a given bandwidth, e.g. in ❶, a flat differential phase shift characteristic is achieved by fine tuning of the applied bias voltage V_b, i.e. from $V_{b,i}$ to $V_{b,j}$. V_{Sat} represents the saturation voltage, at which the highest relative permittivity of the LC layer is reached

of not tunable transmission lines and varactors [Goe09, BR08, Rod+94, Poz11]. In the following section, these two classes are discussed further in details. In particular, design method and parameters affecting the device performance are studied.

4.1.1 Transmission Line Variable Delay Lines

Design and fabrication of a transmission line type LC delay lines is a straight forward approach since the LC material itself is utilized as the dielectric of the line. Planar transmission lines such as grounded coplanar waveguides (GCPWs) [Gar+12] and inverted microstrip lines (IMSL) [KFN02, Mue07, Dol+93] have already been presented. Among them, the IMSLs have been studied in the literature intensively because of its inherently low insertion loss. However, none of these studies discussed how to use available material tunability τ_{LC} efficiently. Only in [Goe09], this point is addressed for metal-insulator-metal waveguides, where some of the results are implicitly applicable for the IMSL topology.

In this context, a tuning efficiency η_τ of a tunable LC device is defined by

$$\eta_\tau = \frac{\tau_{\mathrm{eff}}}{\tau_{\mathrm{LC}}} \tag{4.1}$$

as a ratio of effective device tunability τ_{eff} to material tunability τ_{LC}. The tuning efficiency shows quantitatively how much of the material tunability is utilized by a proposed device layout. For the transmission lines, τ_{eff} is calculated by

$$\tau_{\mathrm{eff}} = \frac{\varepsilon_{r,\mathrm{eff},\parallel} - \varepsilon_{r,\mathrm{eff},\perp}}{\varepsilon_{r,\mathrm{eff},\parallel}}, \tag{4.2}$$

where $\varepsilon_{r,\mathrm{eff},\parallel}$ and $\varepsilon_{r,\mathrm{eff},\perp}$ are the effective relative permittivities when the LC molecules are aligned parallel and perpendicular, respectively, to the radio frequency (RF) signal. η_τ is an essential requirement related to the compactness of the delay line in such a way that the higher η_τ, the lower the IMSL physical length. For the following investigations, τ_{LC} is always assumed to be 0.25.

The parameters affecting η_τ are the height of the LC layer h_{LC}, permittivity $\varepsilon_{r,\mathrm{Sub}}$ and height h_{Sub} of the carrier substrate, where the signal electrode is implemented in, and the width w of the signal electrode. For the analysis, h_{LC} is considered first, since a higher h_{LC} results in higher response times (see Sect. 2.2.2) and a lower one causes high metallic losses [Goe09]. Figure 4.2 shows normalized insertion loss IL against the LC layer thickness for different characteristic impedances of the IMSL. These values are obtained by using the electromagnetic simulation tool Sonnet Software. The metallic and dielectric losses are taken into account in the simulations. The insertion loss can be reduced considerably by increasing h_{LC} and/or decreasing the characteristic impedance Z_{Line} of the IMSL. This is due to the fact that the current density is reduced in both cases by either lowering the field strength or increasing

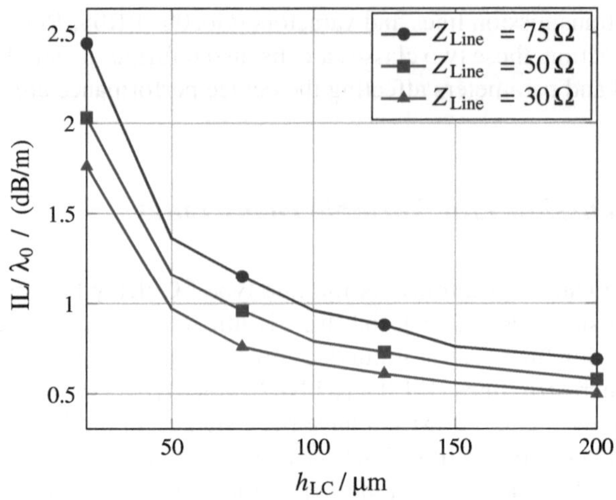

Fig. 4.2 Normalized insertion loss *IL* against the height of the liquid crystal layer h_{LC}. The results are given at 20 GHz when the LC layer and substrate have loss tangents of tan δ 0.006 and 0.008, respectively. Electrodes of the IMSL line are made of 2 μm thick copper with a conductivity of $\sigma_{\text{copper}} = 5.8 \times 10^7$ S/m

the signal width, respectively. Even so, as a compromise between the response time and the insertion loss, h_{LC} of the IMSL is specified to be 100 μm in this work.

According to the formulas given in [Mue07], the physical length of the IMSL should be $4.2\lambda_0$ to obtain a maximum differential phase shift of $\Delta\phi_{\beta,\text{max}} = 360°$ when $\tau_{\text{LC}} = 0.25$ and $\eta_\tau = 1$. Figure 4.2 tells that the insertion loss of this delay line is less than 4 dB when $h_{\text{LC}} = 100$ μm. This leads to a variable delay line efficiency of $\eta_{\text{VDL}} = \Delta\phi_{\beta,\text{max}}/\text{IL} = 90°/\text{dB}$ theoretically. In reality, however, the delay line provides less $\Delta\phi_{\beta,\text{max}}$ since η_τ is less than 1 and therefore, features lower η_{VDL} as will be discussed in the followings.

In Fig. 4.3, the tuning efficiency versus the characteristic impedance Z_{Line} is presented for different carrier substrates. During the simulations, h_{LC} and h_{Sub} are kept to be 100 μm. To calculate τ_{eff} (and so η_τ), $\varepsilon_{r,\text{eff}}$ of the IMSL is determined for each given Z_{Line}, $\varepsilon_{r,\text{Sub}}$. As can be seen in Fig. 4.3, η_τ is maximized when Z_{Line} is lowered or a carrier substrate is utilized with a low relative permittivity. In both cases, the fringing fields, which refer the field propagating along the not tunable carrier substrate, are reduced. The RF signal propagates more in LC layer than in the not tunable substrate, leading to higher η_τ. In applications, it is not always possible to have an arbitrary Z_{Line}, indeed, it is preferred to achieve $Z_{\text{Line}} = 50\,\Omega$ to integrate the delay line into a system. In such a case, η_τ can be increased considerably by choosing a thinner carrier substrate as shown in Fig. 4.4.

Generally, $\eta_\tau < 1$ holds always true for the IMSLs because of the fringing fields. These fields can be lowered by decreasing Z_{Line} and/or by utilizing a thin carrier substrate with a low relative permittivity.

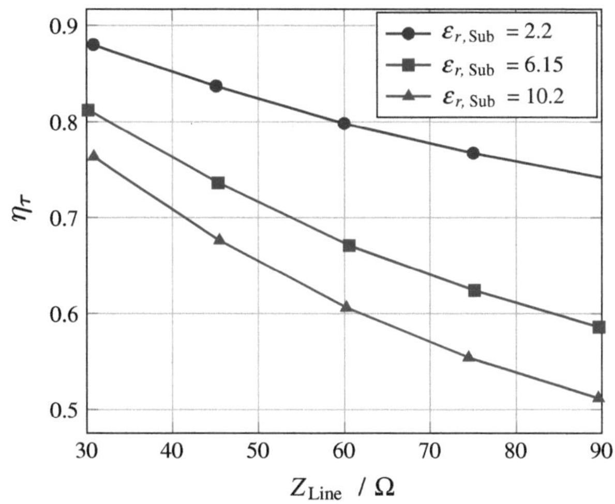

Fig. 4.3 Tuning efficiency η_τ of an inverted microstrip delay line versus its characteristic impedance Z_{Line} for different substrates

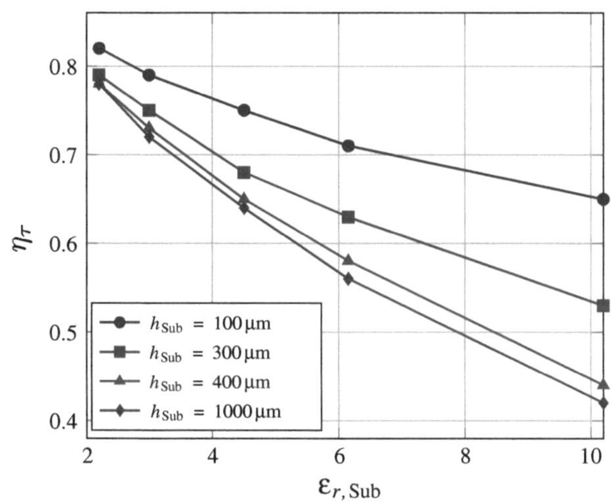

Fig. 4.4 Tuning efficiency η_τ of an inverted microstrip line for different substrates when $Z_{\mathrm{Line}} = 50\,\Omega$

4.1.2 Periodically Loaded Variable Delay Lines

Beside the transmission line topology, an LC based variable delay line can be implemented in periodically loaded line topology. The structure includes *unit elements* (UE), which consists of identical not tunable transmission lines (NTL) loaded with a tunable component. This is illustrated by an infinitely long periodical structure in

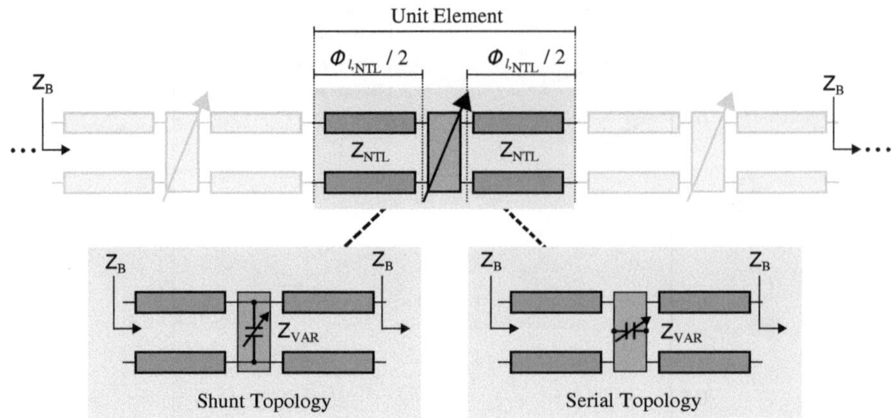

Fig. 4.5 Schematic of a periodically loaded transmission line. *Left* The varactor loads can be connected to parallel to the not tunable transmission lines. *Right* The varactor loads can be connected in serial

Fig. 4.5. The delay operates in either passband or stopband characteristic depending on how the loads are connected to NTLs. This could be in shunt or serial configuration, accordingly.

The loaded line approach is first proposed in [Goe09] for LC devices. There, coplanar waveguide (CPW) is loaded with shunt LC varactors. In the frame of this work, several novel loaded delay line topologies are fabricated and measured. However, before going into the details of these topologies, the serial and shunt configurations are analyzed and their theoretical limits are discussed in the following part.

Loaded Delay Line Architectures: Shunt and Serial Configurations

The analyzes are given for one unit element herein, assuming that the structure includes infinite number of unit elements ($N \rightarrow \infty$). As shown in [Goe09], these analyzes can be carried out even for finite number of unit elements as long as the element number is higher than a minimum number N_{min}, which is defined by the given design requirements (see [Goe09]).

The transmission (ABCD) matrix of the unit element, namely A_{UE}, can be written as [Poz11]

$$A_{\text{UE, Shunt/Serial}} = A_{\text{NTL}} \cdot A_{\text{VAR, Shunt/Serial}} \cdot A_{\text{NTL}}, \tag{4.3}$$

where A_{NTL} is the transmission matrix of the half of a not tunable transmission line and $A_{\text{VAR, Shunt/Serial}}$ is the transmission matrix of the varactor, which is connected to the NTLs in either shunt or serial. For a lossless transmission line, A_{NTL} is given by

$$A_{\mathrm{NTL}} = \begin{pmatrix} \cos\frac{\phi_{\ell,\mathrm{NTL}}}{2} & jZ_{\mathrm{NTL}}\sin\frac{\phi_{\ell,\mathrm{NTL}}}{2} \\ \frac{j}{Z_{\mathrm{NTL}}}\sin\frac{\phi_{\ell,\mathrm{NTL}}}{2} & \cos\frac{\phi_{\ell,\mathrm{NTL}}}{2} \end{pmatrix}. \tag{4.4}$$

In Eq. (4.4), Z_{NTL} and $\phi_{\ell,\mathrm{NTL}}$ are the characteristic impedance and the electrical length of the not tunable transmission line, respectively. $A_{\mathrm{VAR,\ Shunt/Serial}}$ is calculated by

$$A_{\mathrm{VAR,\ Shunt}} = \begin{pmatrix} 1 & 0 \\ \frac{1}{Z_{\mathrm{VAR}}} & 1 \end{pmatrix} \quad \text{or} \quad A_{\mathrm{VAR,\ Serial}} = \begin{pmatrix} 1 & Z_{\mathrm{VAR}} \\ 0 & 1 \end{pmatrix}. \tag{4.5}$$

In Eq. (4.5), Z_{VAR} is the impedance of the varactor that is $Z_{\mathrm{VAR}} = 1/(j\omega C_0)$. ω is angular frequency and C_0 is the capacitance of the varactor, which is the average of the minimum C_{min} and maximum C_{max} capacitances of the varactor. These capacitances are defined by the relative permittivities of the LC material $\varepsilon_{r,\mathrm{LC}}$. The motivation for doing the calculations with the impedances is that the results obtained here are applicable to any operating frequency. For the shunt topology, the corresponding transmission matrix of the unit element $A_{\mathrm{UE,\ Shunt}}$ is obtained as

$$\begin{pmatrix} \cos\phi_{\ell,\mathrm{NTL}} + \frac{jZ_{\mathrm{NTL}}}{2Z_{\mathrm{NTL}}}\sin\phi_{\ell,\mathrm{NTL}} & jZ_{\mathrm{NTL}}\sin\phi_{\ell,\mathrm{NTL}} + \frac{Z_{\mathrm{NTL}}^2}{2Z_{\mathrm{NTL}}}(\cos\phi_{\ell,\mathrm{NTL}} - 1) \\ j\frac{\sin\phi_{\ell,\mathrm{NTL}}}{Z_{\mathrm{NTL}}} + \frac{1}{2Z_{\mathrm{NTL}}}(\cos\phi_{\ell,\mathrm{NTL}} + 1) & \cos\phi_{\ell,\mathrm{NTL}} + \frac{jZ_{\mathrm{NTL}}}{2Z_{\mathrm{NTL}}}\sin\phi_{\ell,\mathrm{NTL}} \end{pmatrix}. \tag{4.6}$$

Similarly, the transmission matrix of the serial topology $A_{\mathrm{UE,\ Serial}}$ is

$$\begin{pmatrix} \cos\phi_{\ell,\mathrm{NTL}} + \frac{jZ_{\mathrm{NTL}}}{2Z_{\mathrm{NTL}}}\sin\phi_{\ell,\mathrm{NTL}} & jZ_{\mathrm{NTL}}\sin\phi_{\ell,\mathrm{NTL}} + \frac{Z_{\mathrm{NTL}}}{2}(\cos\phi_{\ell,\mathrm{NTL}} + 1) \\ j\frac{\sin\phi_{\ell,\mathrm{NTL}}}{Z_{\mathrm{NTL}}} + \frac{Z_{\mathrm{NTL}}}{2Z_{\mathrm{NTL}}^2}(\cos\phi_{\ell,\mathrm{NTL}} - 1) & \cos\phi_{\ell,\mathrm{NTL}} + \frac{jZ_{\mathrm{NTL}}}{2Z_{\mathrm{NTL}}}\sin\phi_{\ell,\mathrm{NTL}} \end{pmatrix}. \tag{4.7}$$

The electrical length ϕ_ℓ and *Bloch Impedance* Z_{B}, which is the characteristic impedance experienced by waves propagating through a periodic structure, are given for per unit length, respectively, by [Poz11]

$$\phi_\ell = \ln\left(A_{\mathrm{UE},11} + \sqrt{A_{\mathrm{UE},11}^2 - 1}\right) \tag{4.8}$$

and

$$Z_{\mathrm{B}} = \sqrt{A_{\mathrm{UE},12}/A_{\mathrm{UE},21}}. \tag{4.9}$$

During the following investigations, the Bloch impedance is specified to be $Z_{\mathrm{B}} = 50\,\Omega$ as being a practical requirement. A corresponding varactor impedance is calculated for a given Z_{NTL} and a $\phi_{\ell,\mathrm{NTL}}$, which results in $Z_{\mathrm{B}} = 50\,\Omega$. The numerical computing environment MATLAB (from MathWorks) is utilized for these calculations. $\phi_{\ell,\mathrm{NTL}}$ is limited to be in a range of 0 to π to have unique solutions. The calculated Z_{VAR} are given in Fig. 4.6 for the both shunt and serial topologies. As Z_{NTL} and/or $\phi_{\ell,\mathrm{NTL}}$ increases, the absolute values of $Z_{\mathrm{VAR,\ Shunt}}$ decrease whereas those of $Z_{\mathrm{VAR,\ Serial}}$ increase. This means that the desired C_0 values increase for the

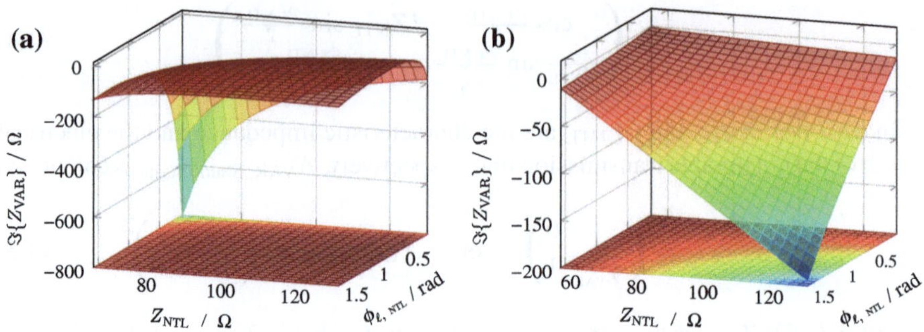

Fig. 4.6 Calculated varactor impadences Z_{VAR} for **a** shunt and **b** serial topologies. The characteristic impedance Z_{NTL} and the electrical length $\phi_{\ell,\mathrm{NTL}}$ of not tunable lines are specified. The corresponding Z_{VAR} is calculated, which leads to $Z_{\mathrm{B}} = 50\,\Omega$

shunt topology and decrease for the serial topology. From the technological point view, it could be a challenge to design high capacitances for the shunt topology. Any additional parasitic inductor reduces the self-resonant frequency of the varactor since a high capacitance is desired.

To determine the maximum differential phase shift $\Delta\phi_{\beta,\mathrm{max}}$, two electrical lengths of the resultant unit elements are calculated when the varactor's capacitance is equal to C_{min} and C_{max}. The difference of these two electrical lengths gives $\Delta\phi_{\beta,\mathrm{max}}$ originated from the varactor tuning. It should be reminded that Z_{VAR} is calculated when the varactor's capacitance is equal to C_0 that is $C_0 = (C_{\mathrm{max}} + C_{\mathrm{min}})/2$. $Z_{\mathrm{VAR,\,min}}$, which is the impedance of the varactor when C_{min} is affective, is calculated by $Z_{\mathrm{VAR}}(2 - \tau_{\mathrm{LC}})/(2 - 2\tau_{\mathrm{LC}})$ for a given material tunability τ_{LC}. Here, it is assumed that the capacitance is proportional to the material tunability and therefore, $\eta_{\tau} = 1$. This is a reasonable assumption since the theoretical maximums of the shunt and serial topologies are under investigation. Similarly, $Z_{\mathrm{VAR,\,max}}$ that is the impedance corresponding to C_{max} is determined by $Z_{\mathrm{VAR}}(1 - \tau_{\mathrm{LC}}/2)$. In a Matlab script, two electrical lengths are calculated for $Z_{\mathrm{VAR,\,min}}$ and $Z_{\mathrm{VAR,\,max}}$, respectively, for given Z_{NTL} and $\phi_{\ell,\mathrm{NTL}}$. The corresponding $\Delta\phi_{\beta,\mathrm{max}}$ are given in Fig. 4.7a–b for different LC tunabilities. For comparison of the two topologies, $\Delta\phi_{\beta,\mathrm{max}}$ of the shunt unit element are normalized by those of the serial one. These results are plotted in Fig. 4.7c. Several interpretations can be done from these results:

i. For a given τ_{LC}, $\Delta\phi_{\beta,\mathrm{max}}$ of the two topologies increase as Z_{NTL} and $\phi_{\ell,\mathrm{NTL}}$ increase. This is essential because for a desired $\Delta\phi_{\beta}$ that is $\Delta\phi_{\beta} < \Delta\phi_{\beta,\mathrm{max}}$, it is not necessary to maximize both Z_{NTL} and $\phi_{\ell,\mathrm{NTL}}$ simultaneously. Indeed, only Z_{NTL} can be maximized to keep $\phi_{\ell,\mathrm{NTL}}$ as low as possible. This leads to a shorter delay line [Goe09] since $\phi_{\ell,\mathrm{NTL}}$ is proportional to the physical length of the not tunable line.
ii. The shunt topology provides more differential phase shift than the serial one. This can be explained by means of equivalent circuits of the unit elements given in Fig. 4.8. On the one hand, the not tunable transmission lines of the unit elements

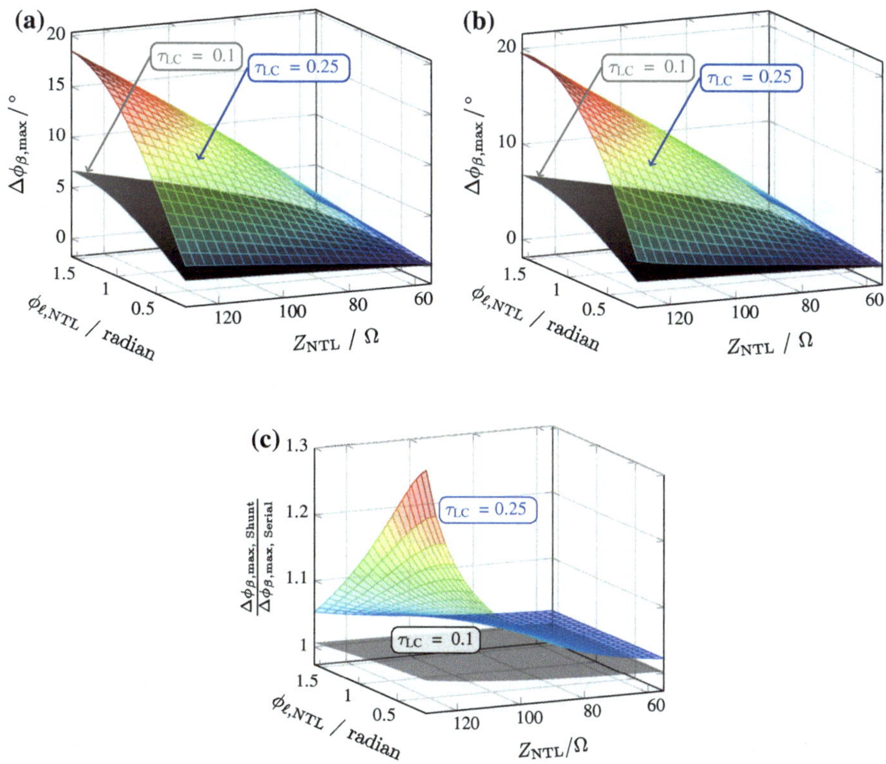

Fig. 4.7 Maximum differential phase shift $\Delta\phi_{\beta,\max}$ provided by **a** shunt and **b** serial unit elements. During the calculations Z_{VAR} are determined for given Z_{NTL} and $\phi_{\ell,NTL}$, resulting in $Z_B = 50\,\Omega$. **c** $\Delta\phi_{\beta,\max}$ of the shunt unit element is normalized by those of the serial unit element

are represented by a serial inductor $Z_{NTL,\ IND}$ and a shunt capacitor $Z_{NTL,\ CAP}$. On the other hand, the equivalent circuit of the resultant loaded line is given by a serial inductor $Z_{TL,\ IND}$ and a shunt effective load $Z_{eff,\ VAR}$. This effective load is the tunable component, where the Z_{VAR} is compromised.

By doing simple circuit mathematics, it can be shown that the effective load impedance of the shunt topology is

$$\frac{1}{Z_{eff,\ VAR,\ Shunt}} = \frac{1}{Z_{NTL,\ CAP}} + \frac{1}{Z_{VAR}} + \frac{1}{\frac{Z_{NTL,\ CAP}\cdot Z_{B,\ Out}}{Z_{NTL,\ CAP}+Z_{B,\ Out}} + Z_{NTL,\ IND}} - \frac{1}{Z_{B,\ Out}}$$
$$= \frac{1}{Z_{NTL,\ CAP}} + \frac{1}{Z_{VAR}} + \frac{1}{Z'_{Shunt}}. \tag{4.10}$$

Similarly, $Z_{eff,\ VAR,\ Serial}$ can be written as

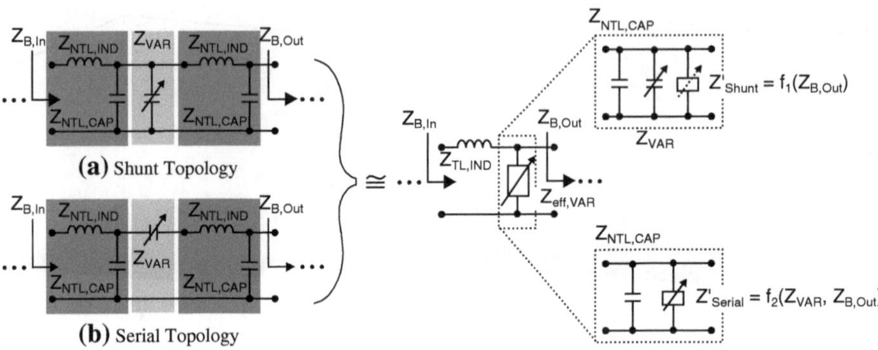

Fig. 4.8 *Left* Equivalent circuits of the shunt and serially loaded unit elements. *Right* The simplified equivalent circuits. $Z_{\mathrm{B,\,In}} = Z_{\mathrm{B,\,Out}} = Z_{\mathrm{B}}$, i.e. $Z_{\mathrm{B}} = 50\,\Omega$

$$\frac{1}{Z_{\mathrm{eff,\,VAR,\,Serial}}} = \frac{1}{Z_{\mathrm{NTL,\,CAP}}} + \frac{1}{\frac{Z_{\mathrm{NTL,\,CAP}} \cdot Z_{\mathrm{B,\,Out}}}{Z_{\mathrm{NTL,\,CAP}} + Z_{\mathrm{B,\,Out}}} + Z_{\mathrm{NTL,\,IND}} + Z_{\mathrm{VAR}}} - \frac{1}{Z_{\mathrm{B,\,Out}}}$$
$$= \frac{1}{Z_{\mathrm{NTL,\,CAP}}} + \frac{1}{Z'_{\mathrm{Serial}}}. \tag{4.11}$$

for the serial topology. As can be seen in Eq. (4.10), the varactor is effective explicitly for the shunt topology by the term $1/Z_{\mathrm{VAR}}$ (see also Fig. 4.8). For the serial topology, on the other hand, the varactor tunability can be seen only in $1/Z'_{\mathrm{Serial}}$ term of Eq. (4.11). This means that the Z_{VAR} term itself does not appear in Eq. (4.11) alone but with some parallel combination to other components. Hence, the varactor tunability is utilized implicitly in the serial topology, leading to less efficiency compared to the shunt one. These analysis are given under the assumption of $Z_{\mathrm{B}} = 50\,\Omega$. For different conditions, which focus more on the theory, one may show that the serial topology provides more $\Delta\phi_{\beta,\max}$ than the shunt one under the given conditions.

iii. When the LC material is tuned, the capacitance of the varactor and therefore, the corresponding Z_{VAR} changes. As a result of this, $Z_{\mathrm{B,\,Out}}$ is affected, which can be seen in Eq. (4.9). The change in Z_{VAR} is desired whereas that of in $Z_{\mathrm{B,\,Out}}$ should be as less as possible. Otherwise, the delay line would be mismatched to the system depending on the tuning state. Specifically, in LC technology, effects of detuned $Z_{\mathrm{B,\,Out}}$ can be neglected because the tunability of microwave LC materials τ_{LC} are less than 0.25. This is actually enough for many applications but less compared to those of ferroelectrics or semiconductors.

iv. $\Delta\phi_{\beta,\max}$ of a shunt unit element is about 20° as shown in Fig. 4.7 when Z_{NTL} and $\phi_{\ell,\mathrm{NTL}}$ are maximized. In practice, this will be less depending on the tuning efficiency of the varactors and obtainable Z_{NTL} and/or $\phi_{\ell,\mathrm{NTL}}$ values of a chosen transmission line technology.

Insertion Loss of a Loaded Delay Line

Insertion losses of a loaded delay line can be distinguished into two. These are the losses caused by the not tunable transmission line and those originated from the varactor. [Goe09] presented obtainable delay line efficiency η_{VDL} depending on these two types of insertion losses. It was shown that η_{VDL} can be increased up to 300°/dB when the quality factors of NTLs and the varactor loads are equal to 10^4 and 200, respectively. However, such a high quality factor of 10^4 is not feasible for planar (printed) transmission lines such as CPW, slotline and microstrip lines. Indeed, their quality factors are less than 250[1] [Mal00].

In the frame of this work, three planar transmission lines are studied as not tunable transmission lines since they enable design of a low profile antenna. These are coplanar waveguide [Sim01], slotline and microstrip line [Gup+96]. They are preferred for several different applications because of their easy fabrication and allowing a designer to mount shunt or serial loads [Sim01]. There are three factors resulting in transmission line losses: Radiation, dielectric and conductor losses. The radiation losses have to be reduced for a transmission line by either optimizing the line geometry or providing an electromagnetic shield [Sim01]. Thus, they can be neglected for a proper design. The dielectric losses are due to the loss tangents of carrier substrates. To reduce this factor, several glass materials have been characterized in Sect. 3.2 to figure out low-loss glasses. The last factor that is the conductor losses depend on the chosen transmission line technology, limited conductivity and surface roughness of the electrodes. Generally speaking, a microstrip line features less insertion loss compared to a CPW o r a slotline [Mal00]. However, the factors affecting the losses are quite complex and depends on a chosen technology, material properties and operating frequency. For instance, CPWs provide less insertion losses than the microstrip lines in CMOS technology [Kad+09] because thin dielectric layers in a CMOS die is not convenient for a microstrip line design. For further details about the loss mechanisms, one can refer to Appendix A.3.2, A.3.3 and [MV09], accordingly for CPW, microstrip and slotline. Therefore, these transmission lines are not compared in a general purpose herein; indeed, their losses will be discussed later, after presenting each loaded delay line topologies in the next sections.

The second source of the losses that is caused by the varactors are considered by a quality factor Q_{VAR} of the varactors. Similar to the line losses, Q_{VAR} is defined by the varactor topology and so, this will be discussed in details in the following sections for each proposed varactor topology.

---

[1] This value can be exemplified by a 50 Ω microstrip line, which is realized on an 0.18 mm thick fused silica with a 2 μm thick copper electrode. It features about 0.1 dB/cm insertion loss at 20 GHz.

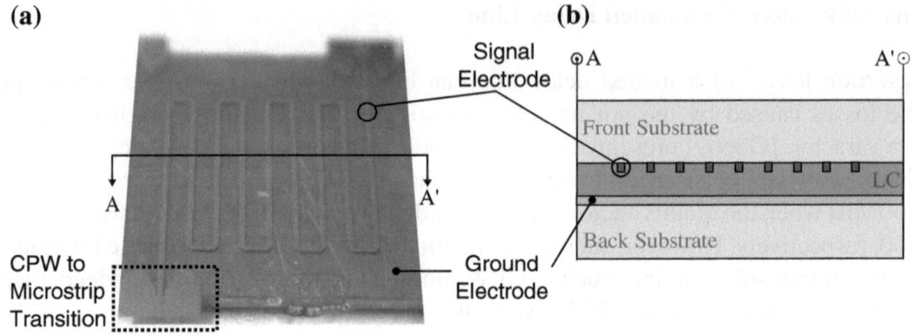

Fig. 4.9 **a** A photograph and **b** cross-sectional view of the realized inverted microstrip delay line

4.2 Realization and Measurements of Variable Delay Lines

4.2.1 Meandered Inverted Microstrip Line

In the light of the previous investigations, an inverted microstrip delay line is realized and measured. As mentioned before, the physical length of the IMSL should be in the order of $4.2\lambda_0$ to provide a high differential phase shift, depending on τ_{LC} and η_τ. Such a long delay line is not feasible for a 2D beam steering phased array. According to the antenna theory [Bal05], an optimum element spacing between the adjacent radiating elements of an array is limited to $0.5\lambda_0$ to $0.65\lambda_0$ to prevent grating lobes. Therefore, a long IMSL has to be implemented into a limited area, which is less than $0.65\lambda_0 \times 0.65\lambda_0$.

One possibility is to meander the delay line [KFN02], enabling to design a compact device. A photograph of such a delay line is shown in Fig. 4.9. The fabrication and operating principle of the structure is discussed in detail in Sect. 2.2. The IMSL is fabricated on the BF33 type glass [SCHc]. Underlying design parameters are given in Table 4.1. The characteristic impedance of the delay line is 50 Ω and its dimensional details are given in Appendix A.4.2.

The structure also includes CPW to microstrip line transitions [Ras+00] at the input and output ports. The transitions are only used for the characterization purpose of the delay line with on-wafer ground-signal-ground (GSG) probes. The delay line is 75 mm long and utilizes an area of about $0.6\lambda_0 \times 0.6\lambda_0$ without the transitions.

Table 4.1 Design parameters of the inverted microstrip line

	Type	ε_r	$\tan\delta$	Thickness
Glass substrate	BF33	4.53	0.008	700 μm
LC layer	TUD-566	2.32 to 3.11	0.0066 to 0.0021	100 μm

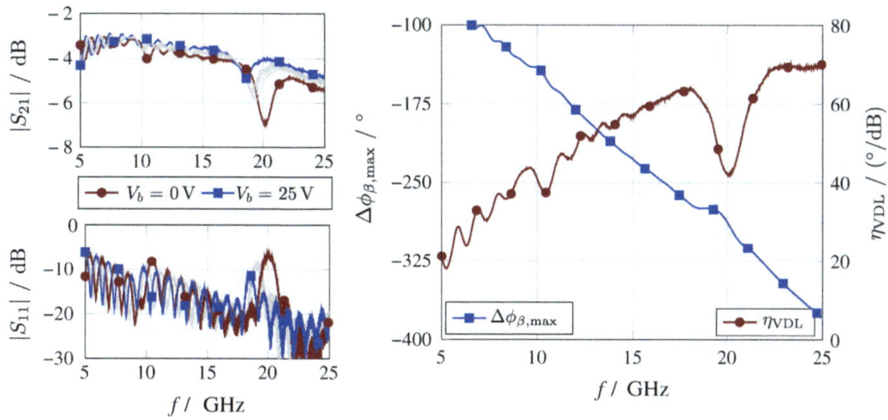

Fig. 4.10 *Left* Magnitudes of the measured scattering parameters of the inverted microstrip line versus frequency f. *Right* The resultant maximum differential phase shift $\Delta\phi_{\beta,\max}$ and delay line efficiency η_{VDL} (*right*)

The measured scattering parameters of the realized IMSL are given in Fig. 4.10. During the measurements, a dc-biasing voltage V_b, which is in the range of 0–40 V, is applied in step of 1 V across the delay line signal and ground electrodes. The electrical characteristics of the delay line remain constant for the bias voltages higher than 25 V. The average IL is about 4 dB. Meanwhile, the measured return loss is greater than 15 dB over the whole frequency range of interest between 15 and 18 GHz and all bias voltages.

The maximum differential phase shift $\Delta\phi_{\beta,\max}$ and the delay line efficiency η_{VDL} are plotted in Fig. 4.10 as well. Based on Sonnet simulations, $\varepsilon_{\mathrm{r,eff},\parallel}$ and $\varepsilon_{\mathrm{r,eff},\perp}$ are determined to be 3.56 and 2.96, respectively. This leads to a tuning efficiency of $\eta_\tau = 0.674$ (see Eq. (4.1)). $\Delta\phi_{\beta,\max}$ is calculated as $-261°$ at 17.5 GHz by substituting the determined $\varepsilon_{r,\mathrm{eff}}$ values in Eq. (4.2). The calculated $\Delta\phi_{\beta,\max}$ matches quite well to the measured one that is $-277°$ at the same frequency. Meantime, η_{VDL} is determined to be 62.5°/dB, including the losses from the CPW to microstrip line transitions. Impacts of the transitions cannot be excluded by using a thru-reflect-line (TRL) calibration technique [CS07] because even the calibration standards have to be filled with the tunable LC material. This leads to imponderable uncertainties in the calibration standards, especially because of a tunable material is utilized. Nevertheless, simulations confirm that the insertion loss of such a transition is about 0.5 dB at 17.5 GHz. When the transitions are excluded by performing a post-processing, IL is about 3.3 dB. This validates the results given in Fig. 4.2 for the 50 Ω IMSL. The efficiency of the line is expected to be higher than 80°/dB at the given operating frequency.

4.2.2 LC Varactor Loaded Coplanar Waveguide

In [Goe09], the shunt LC varactor was implemented by patch as a floating electrode bridging the signal electrode and the ground planes. This topology results in a tuning efficiency of $\eta_\tau = 0.85$ with a varactor quality factor of $Q_{\mathrm{VAR}} = 50$ at the operating frequency of $f = 20\,\mathrm{GHz}$. Here, η_τ is calculated by assuming that there is no fringing field and so, all capacitance is made up by the LC varactor. Then, one can calculate the effective tunability as $\tau_{\mathrm{eff}} = (C_{\max} - C_{\min})/C_{\max}$. In the frame of this work, such a high η_τ is obtained with at least three times higher quality factor at the same operating frequency by means of a novel geometry.

LC Varactor with High Q-Factor

Figure 4.11 shows schematics of the proposed shunt varactor for CPWs, which is called *V-CPW*. The LC material is contained between two glass substrates. The electrodes are implemented in the inner side of the cell, contacting the LC layer. V-CPW includes vertical interconnect accesses (vias) through the LC layer. The patch on the bottom side of the front substrate are connected to the ground electrodes on the top side of the back substrate by using via pillars. The overlapping area, where the signal electrode underlies the bridge, that is $L_{\mathrm{Overlap}} \times W_S$, forms a parallel plate like varactor.

The geometry of the metal bridge connecting the two ground planes is significant to increase Q_{VAR}. A long or narrow bridge decreases the quality factor since a parasitic inductive transmission line segment is introduced between the vias and the varactor. A wider bridge results in a large overlapping area, leading to a high capacitance and

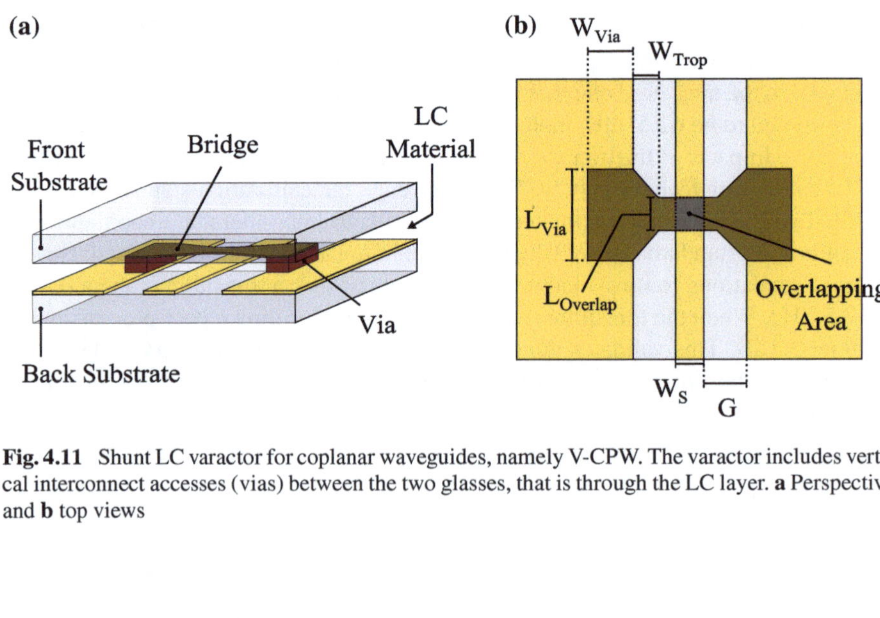

Fig. 4.11 Shunt LC varactor for coplanar waveguides, namely V-CPW. The varactor includes vertical interconnect accesses (vias) between the two glasses, that is through the LC layer. **a** Perspective and **b** top views

a low self-resonant frequency. As a compromise, a trapezoidal geometry is chosen for both ends of the bridge. Additionally, the proposed geometry enables the vias to be implemented with large surface areas to reduce the resistance caused by the vias.

In the design of V-CPW, the width of the signal electrode W_S and the gap between the signal and ground electrodes G are specified for a given Z_{NTL}. In the next step, L_{Overlap} is determined such that the overlapping results in a desired capacitance C_0. For this purpose, full-wave simulations are necessary to obtain the scattering parameters. These parameters are converted into transmission matrixes [Fri94]. Then, according to Eq. (4.5), the capacitance and Q_{VAR} of the shunt varactor are calculated by

$$C_0 = \frac{\Im(A_{\mathrm{VAR,\,Shunt,21}})}{\omega} \tag{4.12}$$

and

$$Q_{\mathrm{VAR}} = \frac{\Im(A_{\mathrm{VAR,\,Shunt,21}})}{\Re(A_{\mathrm{VAR,\,Shunt,21}})} \tag{4.13}$$

respectively, at a given biasing state. The remaining dimensions, which are the length and width of the vias, L_{Via} and W_{Via}, respectively, and the width of the trapezoid W_{Trop}, are specified to maximize η_τ and Q_{VAR}. In Fig. 4.12, results of a parameter study of L_{Via} are presented for different values of W_{Via}. These results are obtained by using Sonnet Software. The simulated varactor consists of two glass substrates, which are 300 μm thick fused silica (see Sect. 3.2). Height of the LC layer is 5 μm that is from the signal electrode to the bridge (metal to metal). During the simulations, $\varepsilon_{r,\mathrm{LC}}$ is tuned from 2.4 to 3.2, resulting in C_{min} and C_{max}, accordingly. As can be seen from the Fig. 4.12, it is not possible to increase both η_τ and Q_{VAR} simultaneously. As the operating frequency increases, getting close to the self-resonant frequency f_r of the varactor, Q_{VAR} reduces. At the same time, the capacitance increases nonlinearly, leading to a high η_τ. Q_{VAR} can be improved considerably by increasing L_{Via} since the parasitic inductances between the vias and the varactor are suppressed. On the other hand, no significant effects of W_{Via} on η_τ and Q_{VAR} are observed. This is most probably due to low, actually negligible, loss of the gold vias. Otherwise, for lossy vias, it is expected to increase Q_{VAR} by increasing W_{Via} as mentioned before since the vias' surface areas increase accordingly.

Figure 4.13 demonstrates the effects of the width of the trapezoid W_{Trop} on η_τ and Q_{VAR}. For these studies, L_{Via} is specified to be a large value, i.e. 250 μm, aiming for further improvement of η_τ and Q_{VAR} obtained from the previous analyzes. A higher W_{Trop} leads to a longer taper and therefore, improves the transition from the vias to the varactor. Hence, Q_{VAR} increases further. However, the effect of W_{Trop} is not as much as that of L_{Via}.

It should be noted that the frequency axes presented in Figs. 4.12 and 4.13 are not normalized to self-resonant frequency f_r. This is done consciously to show that varactors with $f_r > 45$ GHz and with high Q_{VAR} can be designed, simultaneously, by using the proposed structure.

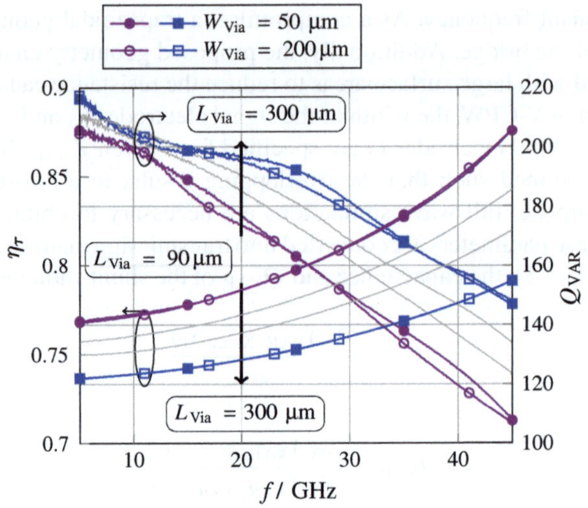

Fig. 4.12 Tuning efficiency η_τ and quality factor of the varactor Q_{VAR} versus frequency for different via lengths L_{Via} and widths W_{Via} when $W_{Trop} = 170\,\mu m$, $W_S = 80\,\mu m$, $G = 170\,\mu m$ and $L_{Overlap} = 90\,\mu m$. The electrodes are made of 2 μm thick gold metal

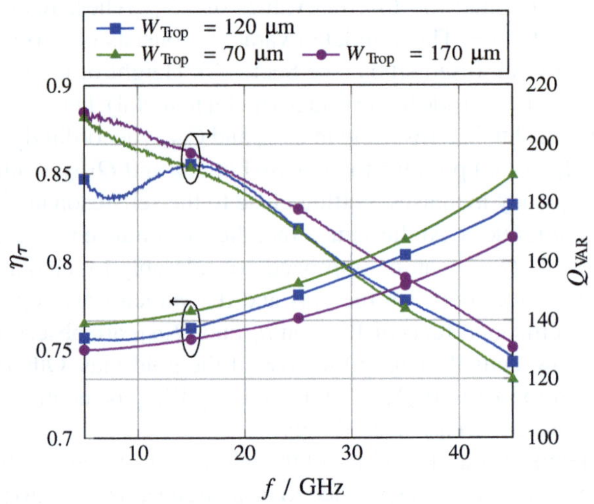

Fig. 4.13 Tuning efficiency η_τ and varactor Q-factor Q_{VAR} versus frequency for different trapezoid widths W_{Trop}. The remaining dimensions are $L_{Via} = W_{Via} = 250\,\mu m$, $W_S = 80\,\mu m$, $G = 170\,\mu m$ and $L_{Overlap} = 90\,\mu m$. The electrodes are made of 2 μm thick gold metal

Realization and Measurement Results of the V-CPW

The theoretical results are verified by measurement results of the proposed V-CPW. The fabrication processes are described in Appendix A.1.1. The realized structure and

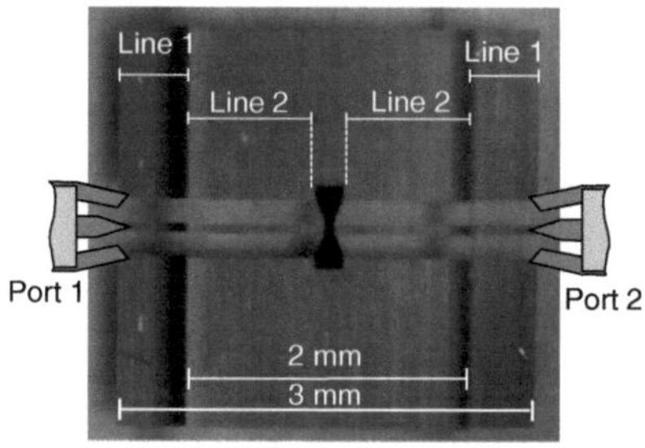

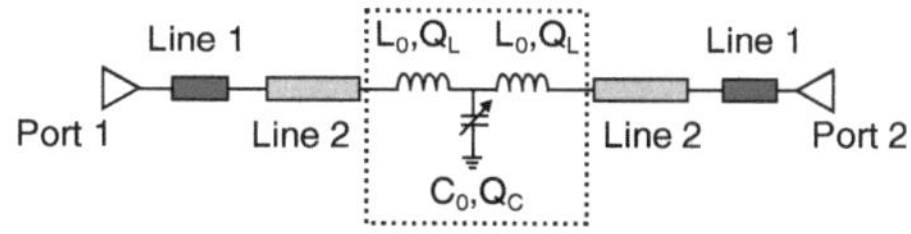

Fig. 4.14 *Left* A photograph of the realized V-CPW. *Right* Its equivalent circuit

Table 4.2 Dimensions of the realized V-CPW varactor

L_{Via}	L_{Overlap}	W_{Via}	W_{Trop}	W_S	G
220 μm	90 μm	100 μm	170 μm	80 μm	180 μm

its equivalent circuit are shown in Fig. 4.14. The height of the LC layer is targeted to be 5 μm and is measured as 5.5 μm by using a confocal microscope. The dimensions of the realized layout are given in Table 4.2. The structure is filled with TUD-566 type LC material. The electrodes are coated by the polyimide layer to anchor the LC molecules parallel to the electrodes. The varactor is connected to the input and output ports through two different types of CPW line segments. The first line, namely *Line 1*, is a 0.5 mm long conventional CPW line. The second one that is *Line 2* is 0.89 mm long and it is sandwiched by two glass substrates.

Scattering parameters of the structure are measured by using an on-wafer measurement station. During the measurements, the LC molecules are aligned by applying a dc bias voltage through a bias tee across the signal and ground electrodes. The applied voltages are in a range of 0–30 V. These results are presented in Fig. 4.15 for $V_b = 0\,\text{V}$ and $V_b = 30\,\text{V}$, resulting to $\varepsilon_{r,\text{LC}} = 2.4$ and $\varepsilon_{r,\text{LC}} = 3.2$, respectively. To evaluate V-CPW, its equivalent circuit model is developed in AWR Microwave Office as given in Fig. 4.14. The shunt varactor is modelled by a T-network, where parasitic inductors L_0 are connected serial to the shunt capacitor C_0. Losses of these lumped elements are considered as well by their own quality factors, which are denoted as Q_L and Q_C, accordingly. Line 1 and line 2 are simulated in Sonnet software and they are included in the circuit model by their corresponding scattering parameter blocks. L_0, Q_L, C_0 and Q_C are determined for two biasing states so, the scattering parameters of the circuit model fits to those obtained by the measurements. The scattering parameters of the equivalent circuit are plotted in Fig. 4.15 by using the parameters given in Table 4.3. Furthermore, the Fig. 4.15 also includes the full wave simulation results of the V-CPW. These three results match well to each other, confirming that the proposed layout enables to design varactors with high Q_{VAR}. According to the parameter fitting, η_τ is calculated as 0.68 and Q_{VAR} is higher than 120.

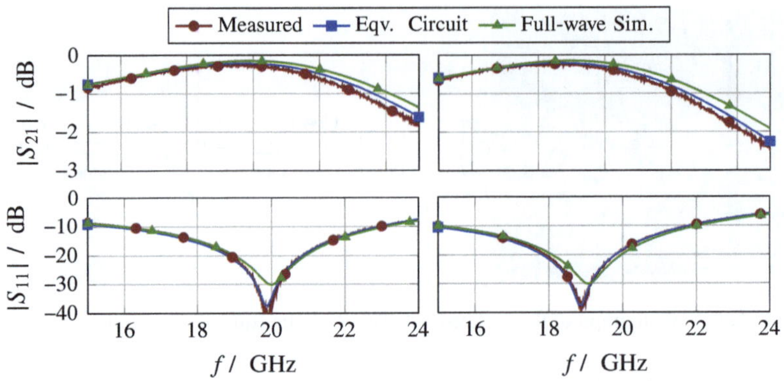

Fig. 4.15 *Left* Magnitudes of the scattering parameters of V-CPW for the not tuned case. *Right* The results are given for the tuned case

Table 4.3 Parameters used for the circuit model of the realized V-CPW

Biasing states	L_0	Q_L	C_0	Q_C
Not tuned state ($V_b = 0\,\text{V}$)	1.8 pH	50	0.0544 pF	120
Tuned state ($V_b = V_{\text{Sat}}$)	1 pH	60	0.0655 pF	150

Additionally, the response times of the varactor are measured and given in Fig. 4.16. Conventionally, the transient response of a capacitor is measured when its capacitance changes between 10 and 90 %. However, in the scope of this works, the response times are defined as the time interval required aligning the LC molecules from perpendicular to parallel or parallel to perpendicular alignments with respect to the electric field of the device (see Sect. 2.2.2). Therefore, here, the response times are recorded under the criteria of change in the capacitance from 0 to 100 % or vice versa. The required time from C_{min} to C_{max} that is t_{on} is measured as 40 ms by applying the bias voltage. When the biased voltage is switched off, the time for obtaining C_{min} is 260 ms. These values can be improved further by employing another LC material, which has lower rotational viscosity than TUD-566 .

Design of Loaded CPW Delay Line

In this section, description and simulation results of a loaded CPW delay line are presented, consisting of the proposed varactors.

The characteristic impedance of the not tunable transmission lines Z_{NTL} is maximized according to geometrical constrains to reduce the physical length of the delay line. These constrains are defined by the electrical characteristics of the chosen glass substrate and the fabrication technology. Similarly, the Bloch impedance Z_{B} is set to be 50 Ω to use the available measurement facilities. For these Z_{NTL} and Z_{B},

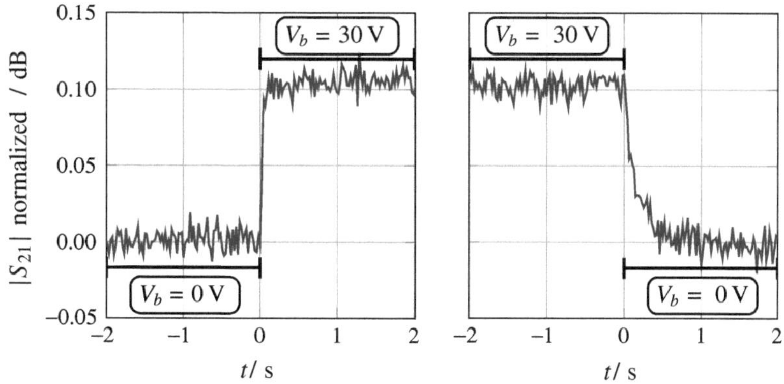

Fig. 4.16 Measured response times of the realized V-CPW

Table 4.4 Design parameters of the loaded CPW delay line

	Type	ε_r	tan δ	Thickness
Glass substrate	BF33	4.53	0.008	700 µm
LC layer	TUD-566	2.32 to 3.11	0.0066 to 0.0021	3 µm
Electrical parameters				
Z_B	Z_{NTL}	$\phi_{\ell,\mathrm{NTL}}$	Z_{VAR}	f_0
40 Ω	100 Ω	0.225 rad	−j32.61 Ω	20 GHz

the required $\phi_{\ell,\mathrm{NTL}}$ and capacitance are calculated by using the formulas given in Sect. 4.1.2. The underlying design parameters are listed in Table 4.4.

In the light of the previous parameter studies of V-CPW, a varactor layout is designed, which provides the desired impedance when $\varepsilon_{r,\mathrm{LC}} = 2.85$. The layout of this varactor is given in Appendix A.4.3. The trapezoidal bridge of the varactor is further optimized to avoid vias. Hence, the fabrication effort as well as the manufacturing costs are reduced. Based on Sonnet simulations, $C_{\mathrm{min}} = 0.209$ pF when $\varepsilon_{r,\mathrm{LC}} = 2.4$ and $C_{\mathrm{max}} = 0.251$ pF when $\varepsilon_{r,\mathrm{LC}} = 3.2$. This leads to a tuning efficiency of $\eta_{\mathcal{T}} = 0.67$, which is almost equal to the one measured for the V-CPW.

Figure 4.17 shows the simulated results of the proposed delay line. The delay line consists of 38 identical unit elements to provide a maximum differential phase shift of 360° at an operating frequency of 20 GHz. Its physical length is 27 mm. The maximum insertion and return losses are about 5.5 dB and 20 dB at f_0. Hence, the delay line offers $\eta_{\mathrm{VDL}} = 65°$/dB at 20 GHz. Because of the fabrication tolerances, one can expect 1 dB more insertion loss after the fabrication, resulting in $\eta_{\mathrm{VDL}} = 55°$/dB. In [Goe09], the delay line was fabricated by using fused silica type glass and provided $\eta_{\mathrm{VDL}} = 60°$/dB at 20 GHz. These results show that, the glass's dielectric losses that is 10 times higher compared to that of the fused silica can be compensated by the high quality factor of the varactor. Therefore, the delay line efficiency is still close to the one presented in [Goe09] even by using low-cost BF33 type glass.

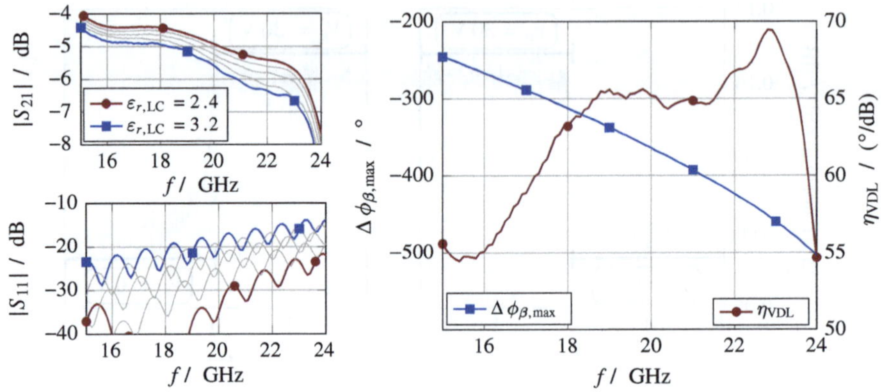

Fig. 4.17 *Left* Magnitudes of the measured scattering parameters of the loaded CPW delay line versus frequency f. *Right* The resultant maximum differential phase shift $\Delta\phi_{\beta,\mathrm{max}}$ and delay line efficiency η_{VDL} (*right*)

4.2.3 LC Varactor Loaded Slotline

In the frame of this work, a novel loaded slotline is studied as a variable delay line. Slotlines are utilized as the not tunable line segments to reduce the impact of the metallic losses. Additionally, because a slotline is used, the varactors can be implemented in parallel plate topology. Hence, not only the insertion loss but also the tuning efficiency of the delay line can be improved by the loaded slot line. Description of the device and its measurement results are presented in this section.

Figure 4.18 shows schematics of an unit of the loaded slotline. The structure consists of stacked glass substrates, where the LC material is encapsulated in between. The electrodes are implemented symmetrically on the top side of the back substrate and on the bottom side of the front substrate. These electrodes include finger-like structures extending from one side. As can be seen in Fig. 4.18b, the finger-like structures on different layers overlap and the overlapping area forms varactors in the parallel plate topology. One of the electrodes of the slotline is elevated with respect to the other one to fill the LC layer in between as shown in Fig. 4.18c. Given that the height of the LC layer h_{LC} is at least 10 times smaller than the width of the slotline W_{Slot}, the elevation of the upper electrode is negligible for the RF propagation.

Selection of the enclosing substrate is significant for the device performance. For the slotline to be practical as a transmission line, radiation must be minimized [Coh69]. This can be accomplished through the use of a high permittivity substrate. However, as the permittivity of the enclosing substrate increases, the tuning efficiency of the varactors decreases. This is due to the fact that the portion of the field going through the not tunable substrate becomes larger and thereby the larger amount of the capacitance is defined by these fringing fields. As a compromise between the radiation loss of the slotline and the tuning efficiency of the varactors, a glass substrate that is AF45 [Af4] is chosen because of its favorite dielectric characteristics.

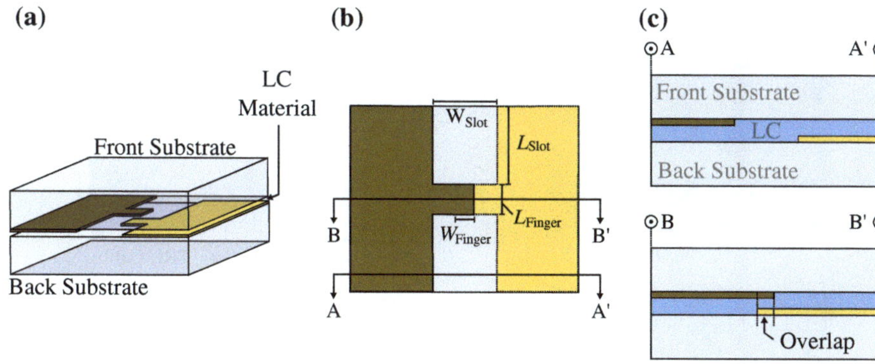

Fig. 4.18 A unit element of the loaded slotline. **a** Perspective, **b** top and **c** cross-sectional views of the unit element

These are $\varepsilon_r = 5.87$ and $\tan\delta = 0.0083$ at 18.2 GHz and the glass thickness is 0.51 mm. Regarding the varactor design, the results presented in Sect. 4.2.2 for the trapezoidal bridge are applicable for the varactors utilized in the loaded slotline.

CPW to Slotline Two-Step Transition

For the on-wafer measurements, a transition from CPW to slot line is required. After studying different topologies [Gup+96, Het+00], a two-step transition is designed because of its broadband operation. In the first step, a CPW to microstrip line transition [Ras+00] is designed, which is then followed by a microstrip to slotline transition in the second step. Such a back-to-back two-step transition is realized and measured to test its RF performance. Photographs of the structure are shown in Fig. 4.19a. The structure is fabricated on RO3206 substrate from Rogers Corporation. Its electrical characteristics ($\varepsilon_r = 6.15$ and $\tan\delta = 0.0027$) are close to those of AF45. The GSG probes touch down on a short CPW line, i.e. $\lambda_0/8$ in length. The RF signal on the CPW is fed through a microstrip line by extending the signal electrode of the CPW. The ground electrodes of the CPW are capacitively coupled to the ground of the microstrip line, which is on the back side of the substrate. Then, the RF signal on the microstrip line is fed to the slotline. Here, the slot line is fabricated on the ground plane of the microstrip line and it is shorted at both ends. The structure is characterized under two different calibrations of the measurement setup. For the first one, a conventional serial, open, line and through (SOLT) calibration is applied. For the second one, effects caused by the transitions are excluded by applying a through, reflect, line (TRL) type of calibration. In this case, the reference plane is moved to the slotline section and the length of the measured slotline is 2 mm. Magnitudes of the measured scattering parameters are plotted in Fig. 4.19b. The insertion loss of the complete structure is about 2–4 dB and this value reduces to 0.5 dB when the TRL

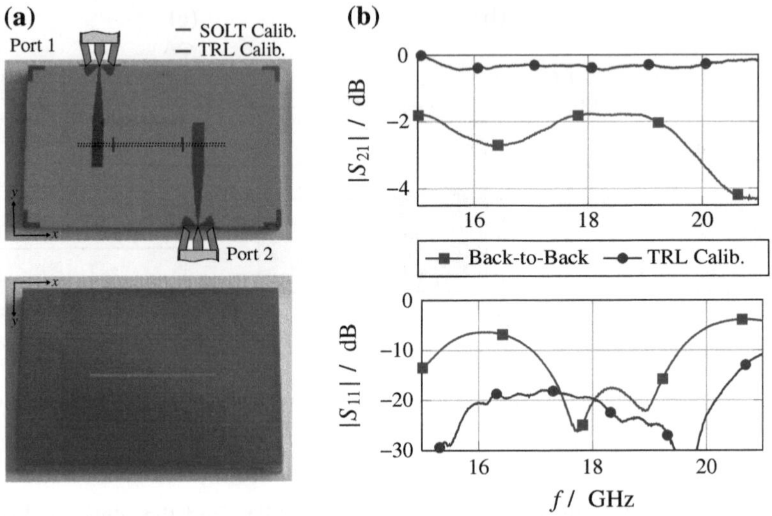

Fig. 4.19 **a** Photographs and **b** measured scattering parameters of the back-to-back CPW to microstrip line two-step transition

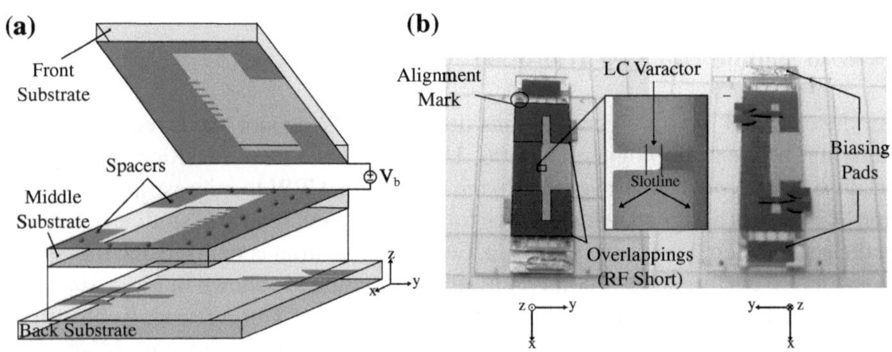

Fig. 4.20 **a** Exploited schematic and **b** photographs of the loaded slotline prototype. The structure includes CPW to slotline transitions and loaded slotline with 8 unit elements. The LC material is filled in between the front and middle substrates

calibration is applied. For both calibrations the reflections are low, for instance, they are less than −20 dB at 18 GHz.

Realization and Measurement Results of the Loaded Slotline

Figure 4.20 shows a schematic and photographs of the realized loaded slotline. The structure is fabricated by using the standard photolithography processes (see Appendix A.1.1). The design parameters of the structure are given in Table 4.5.

Table 4.5 Design parameters of the loaded slotline

	Type	ε_r	$\tan\delta$	Thickness
Glass substrate	AF45	5.87	0.0083	510 μm
LC layer	TUD-566	2.32 to 3.11	0.0066 to 0.0021	3 μm
Electrical parameters				
Z_B	Z_{NTL}	$\phi_{\ell,\text{NTL}}$	Z_{VAR}	f_0
40 Ω	75 Ω	0.131 rad	−j67 Ω	18 GHz

Because of the microstrip to slotline transitions used in the two-step transition, the slotline has to be shorted at both ends. However, it is formed by stacking two electrodes, where one of the electrodes is elevated with respect to the other one (Fig. 4.18) to fill the tunable LC material in between. To obtain the necessary RF short, the electrodes forming the slotline overlap at both ends, where the short termination is needed. This results in a large capacitance and therefore, low impedance.

A cavity is formed between the top and middle substrates by using micropearls from Seikui Chemical Co. LTD. These are made of rigid silica particles and have a diameter of 4 μm. This cavity is filled with TUD-566 type LC. The electrodes are coated by the polyimide layer to anchor the LC molecules parallel to the electrodes. Later on, a desired LC molecule orientation and therefore, the LC layer relative permittivity is accomplished by applying a bias voltage V_b across the two electrodes through the biasing pads (see Fig. 4.20).

The loaded slotline is measured by applying the TRL calibration. The measured scattering parameters are plotted in Fig. 4.21. The insertion and return losses of the loaded line are 1 dB and 15 dB, respectively, at the operating frequency of 18 GHz. The loaded line consists of 8 unit elements and it is 7 mm long, which is equal to $0.42\lambda_0$ at the operating frequency. This small prototype provides a $\Delta\phi_{\beta,\max}$ of 43°, which is close to the simulated one that is ≈50°. The corresponding delay line efficiency is determined to be 42°/dB at 18 GHz from the measurement results.

4.2.4 LC Varactor Loaded Microstrip Line

The efficiency of the loaded variable delay line can be further improved by using a microstrip line for the not tunable transmission line segment. This is due to the fact that the metallic losses of the microstrip line can be reduced considerably if the transmission line is designed properly.

A loaded microstrip delay line can be designed by utilizing the proposed trapezoidal bridge periodically on top of the signal electrode of the microstrip line. This is similar to the V-CPW described in Fig. 4.11, where the parallel plate like varactor is accomplished between the signal and bridge electrodes. However, unlike to the CPW, the electrodes forming the microstrip line are not on the same side of a dielectric.

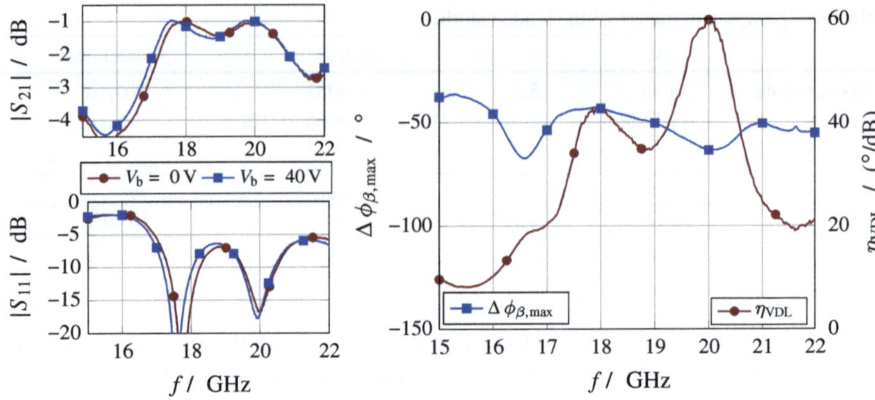

Fig. 4.21 *Left* Magnitudes of the measured scattering parameters of the loaded slotline versus frequency f. *Right* The resultant maximum differential phase shift $\Delta\phi_{\beta,\max}$ and delay line efficiency η_{VDL}

Hence, the bridge has to be connected to the ground electrode by means of vertical interconnect accesses (vias) through the LC layer and through the glass substrate. Vias through the LC layer have already been successfully realized and tested in this work for the realization of V-CPW in Sect. 4.2.2. The vias through the glass substrate are avoided in general since drilling holes to a glass substrate is not a standard technique used in LC display technology [YW06]. This increases the manufacturing costs and may lead to leakage.

The second type of vias can be avoided by using another type of varactor. One possibility is to load the not tunable line with *split rings resonators* (SRRs). SRR has been first proposed in [Pen+99] as a unit element of an artificial material, featuring artificial magnetic dispersion of permeability. The structure consists of split- and concentric-rings, which are made of conducting materials. After the first realization in [Smi+00] as a proof of the concept, SRRs and their complementary structures, so called *complementary split rings resonators* (CSRRs) have been studied intensively for planar metamaterial transmission lines [Cap09]. The electrical sizes of these structures are about one tenth of the free space wavelength λ_0 at the resonance frequency [Mar+04] and therefore, enable compact designs. SRRs can be excited by means of a time-varying magnetic field and exhibit stopband characteristics. Their applications in CPW [Mar+03] and microstrip line [Gil+05] technologies have been presented for compact filter realizations. On the other hand, CSRRs are excited by a time-varying electric field and provide stopband characteristics. In [Fal+04], a microstrip line is loaded with CSRRs. There, the resonators are designed on the ground electrode of the microstrip line, which underlie the signal electrode. Two other types of the resonators are proposed as well to connect them serial to the transmission line's signal electrode. These resonators are so called *open split rings resonator* (OSRR) [Mar+04] and *open complementary split rings resonator* (OCSRR) [DS+09], which are excited by the current flowing on the signal electrode. OSRR and OCSRR based structures have highpass and stopband characteristic, respectively.

Different methods are attempted successfully to tune the electrical characteristics of the planar metamaterial transmission lines. For instance, SRR based devices are loaded with semiconductor varactors [Gil+06, VBM08] and they are implemented on a ferroelectric material, i.e. barium-strontium-titanate [Gil+11]. Both methods are applicable efficiently only for $f < 10\,\mathrm{GHz}$ because of the self-resonant frequencies of the varactors or high dielectric losses of the ferroelectrics at higher microwave frequencies [Gei09]. In the frame of this work, tunable OSRRs are designed in LC technology. These resonators are used as varactors to load a microstrip line. Unlike the loaded lines presented in the previous sections, the microstrip line is loaded serially. Owing to the LC technology and the small electrical size of the OSSRs, compact variable delay lines can be designed by using the proposed layout, operating at higher microwave frequencies ($f > 10\,\mathrm{GHz}$). In the following section, OSRR based loaded microstrip delay line is presented in details.

Description and Design of the LC Based OSRR

The OSRR presented in [Mar+04] consists of two concentric metal rings, which are implemented on the same side of a substrate. The rings are electromagnetically coupled to each other and therefore, a desired capacitance is accomplished between the edges of the rings. As shown in Fig. 4.22, this structure is adapted to the LC technology in such a way that two identical rings are fabricated on the front and back substrates. The rings face to each other and have contact to the LC material, which is inserted between the substrates. The desired capacitance is achieved between the ring electrodes and it can be tuned because of the LC material. The ground electrode of the microstrip line is located on the bottom side of the back substrate. It should be noted that the signal electrode implemented on the bottom side of the front substrate is elevated compared to the one located on the top side of the back substrate. As long as a high aspect ratio that is about 10, is provided between the height of the back substrate h_{Sub} and that of the LC layer h_{LC}, both microstrip lines can be treated to be identical.

The equivalent circuit of the OSRR is given in Fig. 4.22c [Mar+04]. It includes a resistor R_0, a lossy varactor with an average capacitance C_0 and a quality factor Q and an inductor L_0. R_0 and L_0 represent parasitic components between the not tunable lines and the rings. It should be mentioned, that Q_{VAR} used in the previous sections and will be also used later refers to the quality factor of the equivalent circuit. In [Mar+04], a rectangular slot underlying the OSRR was implemented on the ground electrode. It was designed to be larger than OSRR to prevent any electromagnetic coupling between the rings and the ground electrode. In this work, such a slot is not needed because the coupling between the ground electrode and the rings is avoided by choosing the dimensions of the rings and height of the substrates carefully. This is illustrated by a full wave simulation of the proposed structure by using CST Microwave Studio. Figure 4.23 shows a cross-sectional view of the structure, where

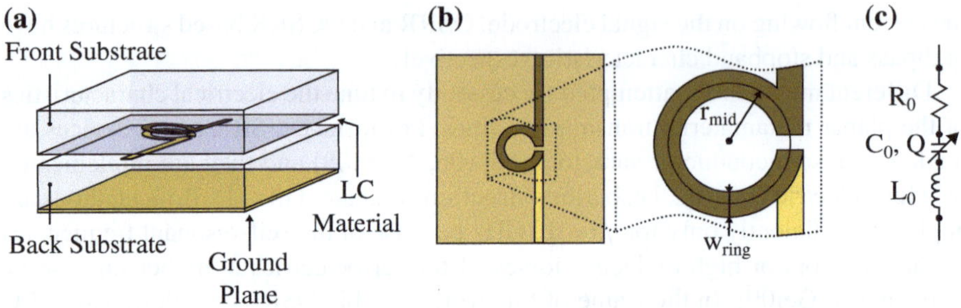

Fig. 4.22 A unit element of the open split rings resonator (OSRR) loaded microstrip variable delay line. **a** Perspective and **b** top views of the unit element. **c** Equivalent circuit of the OSRR

Fig. 4.23 Cross-sectional view of OSRR, where the amplitude of the electric field is plotted at 16 GHz. The colorbar is given in dB scale

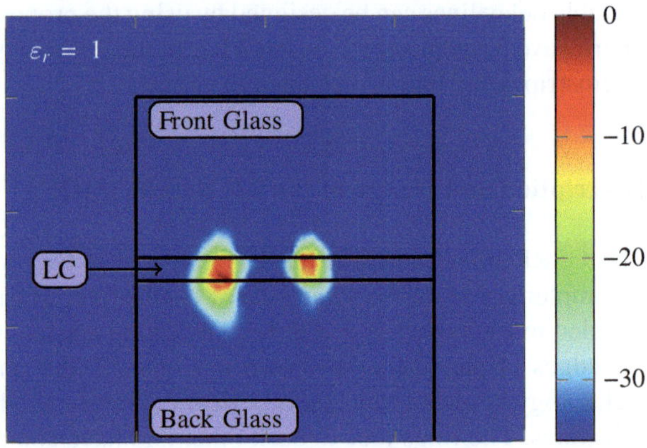

the amplitude of the $\vec{E}$ field is plot at 16 GHz. The view is given through the ring's diameter that is orthogonal to the microstrip lines. The back and front substrates are 700 µm thick BF33 type glass. The aspect ratio of $h_{\text{Sub}}/h_{\text{LC}}$ is equal to 7. The relative permittivity of the LC layer $\varepsilon_{r,\text{LC}}$ is specified to be 2.4 to enhance the field propagating through the back substrate. As can be seen from Fig. 4.23, the RF field is concentrated between the rings and no coupling to the ground electrode is counted. Indeed, the field propagating through the back substrates decreases if h_{LC} is reduced or if $\varepsilon_{r,\text{LC}}$ is increased by applying a bias voltage V_b.

Similar to the V-CPW presented in Sect. 4.2.2, the tuning efficiency η_τ and Q-factor Q_{VAR} of an OSRR based varactor depend on a targeted operating frequency f_0 and varactor's self resonant frequency f_r. η_τ and Q_{VAR} cannot be improved simultaneously. As f_0 approaches to f_r, η_τ increases and Q_{VAR} decreases, accordingly. Hence, the fundamental relations between f_r and the physical dimensions of OSRR have to be known to compromise η_τ and Q_{VAR}. This is the subject of the subsequent paragraphs.

Three dimensions should be specified to design an OSRR layout for a given substrate. These dimensions are the radius of the ring that is assumed to be in the middle of the inner and outer edges of OSRR r_{mid}, width of the rings w_{ring} and the height of the LC layer h_{LC} according to Fig. 4.22b. The self resonant frequency of a serial capacitor is calculated by

$$f_r = \frac{1}{2\pi\sqrt{L_0 C_0}}. \tag{4.14}$$

The capacitance of OSRR C_0 is mainly determined by the overlapping area between the rings since the field strength is maximized at this region (see Fig. 4.23). Similar to a parallel plate capacitor, C_0 is proportional to this area that is $2\pi r_{\text{mid}} w_{\text{ring}}$ when the rings are not split. Hence, the relation between f_r and the ring dimensions can be written as

$$f_r \propto \frac{1}{\sqrt{r_{\text{mid}}}} \quad \text{and} \quad f_r \propto \frac{1}{\sqrt{w_{\text{ring}}}}. \tag{4.15}$$

according to Eq. (4.14). In analogy to the parallel plate capacitor, as h_{LC} increases and the voltage on the rings are kept constant, less electrical charge is stored between the rings since the field strength is reduced. This results in a decrease in C_0, and therefore,

$$f_r \propto \sqrt{h_{\text{LC}}}. \tag{4.16}$$

Figure 4.24 depicts how the ring dimensions affect f_r, verifying the relations given in Eqs. (4.15) and (4.16). Additionally, when the two figures are compared one can see that f_r can be kept constant as the dimensions of the OSRR shrink. Thus, if an LC based OSRR with a low response time (and therefore, low h_{LC}) is designed, the resultant structure will also be more compact compared to the one with high response times. This fact actually holds true for LC based parallel plate like varactors in general and it is shown quantitatively for OSRRs herein.

Realization and Measurement Results of a Loaded Microstrip Delay Line

An OSRR loaded microstrip delay line is designed to operate at 16 GHz by using the formulas given in Sect. 4.1.2. Resultant underlying design parameters are listed in Table 4.6. The layout is fabricated by using the photolithography techniques presented in Sect. 2.2.1. The prototype is shown in Fig. 4.25. It consists of CPW to microstrip line transitions for the measurements and 24 identical unit elements, each including the not tunable microstrip lines and an LC based OSRR as a serial varactor.

The alignment of the LC molecules and therefore, a desired $\varepsilon_{r,\text{LC}}$ is accomplished by using the alignment layer and by applying a bias voltage V_b across the electrodes. The applied bias voltage is delivered to the RF circuitry through resistive bias lines. The bias lines are implemented on the adhesive nickel-chromium layer (Ni-Cr) that is located between the gold electrode and the glass substrate [Saz+11b]. During the fabrication, the Ni-Cr layer is usually etched away. Here, it is covered with a photoresist,

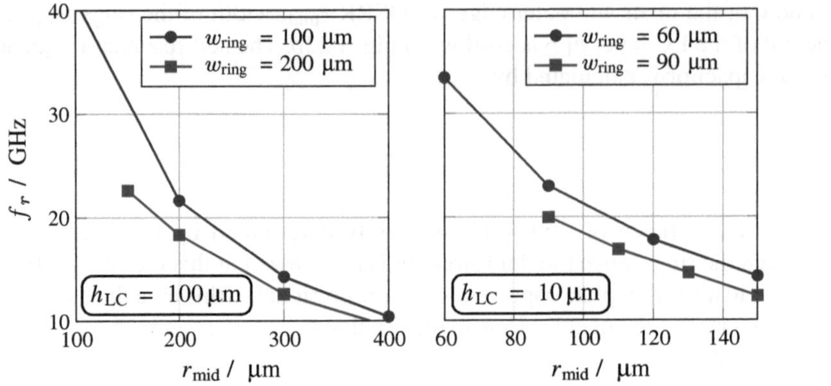

Fig. 4.24 Self resonant frequency f_r of an LC based OSRR varactor versus its physical dimensions. r_{mid} is the radius of the ring and w_{ring} is the width of the ring

Table 4.6 Design parameters of the realized loaded microstrip delay line

	Type	ε_r	$\tan\delta$	Thickness
Glass substrate	BF33	4.53	0.008	700 μm
LC layer	TUD-566	2.32 to 3.11	0.0066 to 0.0021	100 μm
Electrical parameters				
Z_B	Z_{NTL}	$\phi_{\ell,NTL}$	Z_{VAR}	f_0
50 Ω	110 Ω	0.405 rad	−j65.45 Ω	16 GHz

which is then patterned and developed for a desired biasing network. When the Ni-Cr layer is etched, the conductive layer underling the photoresist remains there and

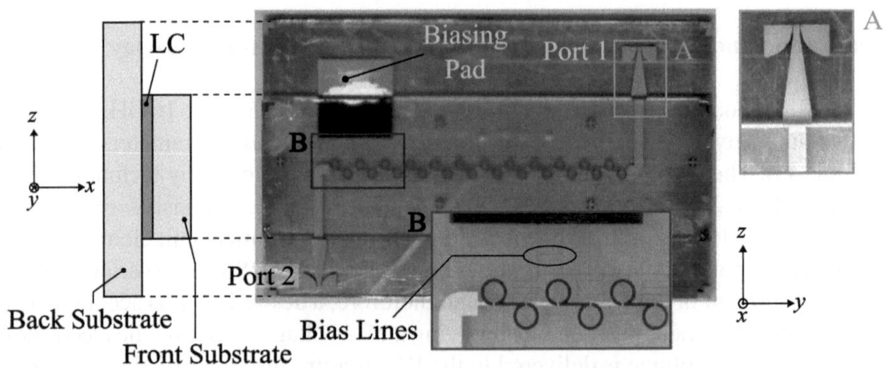

Fig. 4.25 **a** Cross-sectional view and **b** photographs of the realized loaded microstrip delay line. LC based OSRRs are utilized as serial varactors. The structure includes 24 unit elements. The electrodes are made of 2 μm gold metal

forms the biasing network. The drawback is low to moderate resistance of the bias line R_{Line}, which is calculated by

$$R_{\text{Line}} = \frac{1}{\sigma_{\text{Ni-Cr}}} \frac{\ell}{wt}, \tag{4.17}$$

where ℓ and w are the length and width of the bias line, respectively, and t is the thickness of the Ni-Cr layer and $\sigma_{\text{Ni-Cr}}$ is the conductivity of the Ni-Cr layer that is $\sigma_{\text{Ni-Cr}} = 7 \times 10^5$ S/m.

According to ADS simulations, the effects of the bias line on the RF circuitry are negligible if R_{Line} between the two adjacent rings is higher than 10 MΩ. Otherwise, the RF signal leakages and the insertion loss increases. For instance, when R_{Line} is reduced from 10 MΩ to 1 MΩ and 0.1 MΩ, the insertion loss of the unit element increases 0.1 dB and 0.3 dB, respectively. To maintain a high resistance w and t are specified to be 10 μm and 15 nm, accordingly. These are typical minimum limits of the in house facilities. The resultant R_{Line} is then calculated to be around 0.013 MΩ/mm. The bias lines are meandered to increase their overall length. Even then, the resultant length is about 10 mm, leading to 0.13 MΩ bias line resistance. This is not an optimum value and so, have certain affects on the RF performance of the prototype as will be discussed in the next part.

Figure 4.26 shows the measured and simulated scattering parameters of the prototype. The simulations results are given only for zero biasing ($V_b = 0$ V), resulting in $\varepsilon_{r,\text{LC},\perp}$. There are two simulation results, where the first one is obtained when the bias lines are not taken into account. However, when they are considered as in the second simulation results, an increase in the insertion losses is observed over the frequency. The second results match very well to the measured results. The corresponding

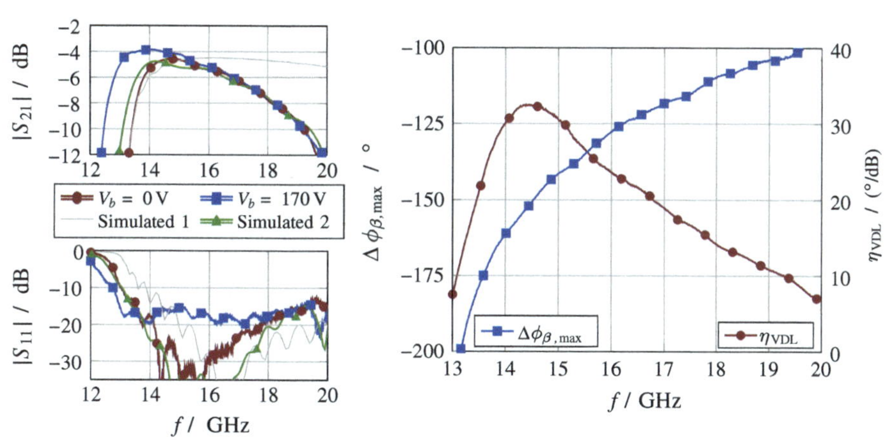

Fig. 4.26 *Left* Magnitudes of the measured scattering parameters of the loaded microstrip delay line versus frequency f. *Right* The resultant maximum differential phase shift $\Delta\phi_{\beta,\text{max}}$ and delay line efficiency η_{VDL} (*right*)

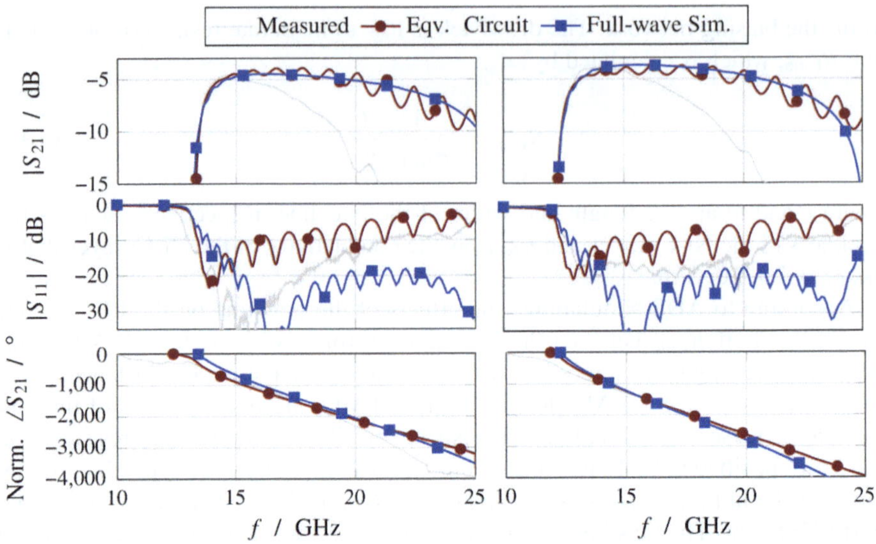

Fig. 4.27 *Left* Scattering parameters of the loaded microstrip delay line for the not tuned case. *Right* The results are given for the tuned case. The abbreviation *Norm.* stands for normalized

Table 4.7 Parameters used for the equivalent circuit of OSRR

Biasing states	R_0	L_0	C_0, Q	$C_{\mathrm{Eqv.}}$	Q_{VAR}
Not tuned state ($V_b = 0\,\mathrm{V}$)	0.5 Ω	1.3 nH	0.0725 pF, 80	0.269 pF	18.74
Tuned state ($V_b = V_{\mathrm{Sat}}$)	0.4 Ω	1.1 nH	0.0950 pF, 100	0.496 pF	11.22

$\Delta\phi_{\beta,\max}$ and η_{VDL} are $-160°$ and 32.5°/dB, respectively, at 14 GHz. The delay line efficiency reduces to 25°/dB at 16 GHz because of the effects of the bias lines.

An equivalent circuit model of the realized variable delay line is developed to evaluate the OSRRs. The microstrip line is simulated in Sonnet software and included in the model with its scattering parameter block. This block is loaded with the equivalent circuit of OSRR (see Fig. 4.22c) to form the unit element. 24 unit elements are cascaded and the microstrip feeding lines are added to both ends. The bias lines are not taken into account because their effect can be suppressed completely by using a low resistive conductor like indium-tin-oxide. Hence, the scattering parameters of the equivalent circuit are fitted into the ones obtained by the full-wave simulations by determining convenient values for the lumped elements in the equivalent circuit. In Fig. 4.27, the scattering parameters of the modeled delay line are given together with the measured and simulated ones. Transmission coefficients of the circuit model and those obtained from the full wave simulations show good agreements. The parameters used for the equivalent circuit of OSRR are listed in Table 4.7. These values have a certain error whereas they are given here to give a basic understanding about the OSRRs performance. Based on these values, equivalent capacitances of OSRR

Table 4.8 Design parameters and measured performances of the variable delay lines

Trans. line	f_0 (GHz)	N	$\ell_{\text{physical}}/\lambda_0$	$\Delta\phi_{\beta,\max}$ (°)	$\text{IL}_{\max}$ (dB)	$\eta_{\text{VDL, Sim}}$ (°/dB)	$\eta_{\text{VDL, Mea}}$ (°/dB)
IMSL	17.5	–	6.126	360	4.12	90	85
Loaded CPW	20	38	1.773	365	5.88	62.1	–
Loaded slotline	18	48	2.399	376	10.50	37.0	42.5
Loaded microstrip	16	40	2.133	373	12.30	29.3	25.0

N refers the unit element number in the *LC* varactor loaded delay lines

$C_{\text{Eqv.}}$ are determined by using Eqs. (4.12) and (4.13). The resultant η_τ is calculated to be 1.83. Such a large value is obtained because the operating frequency is close to the self resonant frequency of OSRR, which is 16.40 GHz at the zero biasing state ($\varepsilon_{r,\text{LC}} = 2.32$) and 15.55 GHz at the full biased state ($\varepsilon_{r,\text{LC}} = 3.11$). At the same time, Q_{VAR} is in a range of 11.22–18.74 for different biasing states. It should be noted that the high losses of OSRR mainly originate from the parasitic resistor and inductor. The LC material losses themselves are less since the quality factor of the lumped capacitor is higher than 80. Therefore, another varactor topology with low or, even negligible, parasitic components is required to improve η_{VDL} of the proposed delay line.

4.2.5 Comparison of Different Topologies

The delay lines presented in the previous parts are compared herein for selected key parameters: Insertion loss, compactness, response time and the reliability. Other essential criteria such as power handling capabilities of the delay lines are not considered since the targeted application is a phased array in a receiving mode. Actually, a phased array antenna can operate either in receiving or transmitting mode since it is a passive and reciprocal device. Having a low insertion loss is more critical for the receiving antenna since the received signal is in a low power level and will be attenuated further by the delay line until it is processed by the electronics.

For a fair comparison, the physical length ℓ_{physical} of the delay lines are increased to achieve $\Delta\phi_{\beta,\max} \approx 360°$ at their operating frequencies f_0. This is confirmed by full-wave simulations. The resultant parameters and performances of these long delay lines are given in Table 4.8. Simulated $\eta_{\text{VDL, Sim}}$ values match quite well to the measured $\eta_{\text{VDL, Mea}}$,[2] which shows the accuracy of the simulations. Slight differences are mainly caused by the fabrication tolerances.

[2] In calculations of $\eta_{\text{VDL, Mea}}$, the insertion losses caused by any transitions are excluded since they are not considered in the simulations as well.

Insertion Loss

The sensitivity of a receiving antenna to weak signals is an essential antenna specification. A convenient measure, which is also proportional to signal-to-noise ratio, is G/T in dB/K [Orb, Sio81, KJDM78]. G is the antenna gain and T is the receiver noise temperature. G/T has to be maximized to improve the system sensitivity. Hence, it is significant for the antenna to prevent the losses caused by the delay lines since they reduce the G and increase the T.

The IMSL provides the highest delay line efficiency η_{VDL} because of the low insertion loss IL of the microstrip line technology. If ILs are normalized by a factor[3] of $\ell_{\mathrm{physical}}/\lambda_0$, the IMLS line has only 0.673 dB insertion loss. Regarding the loaded line topologies, the CPW based structure has the highest η_{VDL} that is 62.1°/dB. This is an interesting result compared to the loaded microstrip line topology since the microstrip line has fewer losses. As was mentioned earlier, there are two mechanisms contributing to the delay line losses. First, the losses caused by the not tunable transmission lines. Figure 4.28 presents the attenuation constants α and the propagation constants β of these lines versus frequency. As it is expected, the microstrip line attenuates the signal less than the CPW. However, unlike to [Ra69], the slotline used in this work has the highest attenuation because of the radiation losses. To obtain the slot-mode, the radiation must be lowered by using a high permittivity substrate [Coh69]. Hence, the field propagating outside the substrates can be suppressed and the energy is confined by the slots with negligible radiation losses [Coh69, Gup+96]. Figure 4.29 shows the amplitude of the E-field of the slotline used in this work. Unfortunately, a high amount of the field propagates in the air region since the relative permittivity of the AF45 type of glass is 6.2. This leads to high radiation losses and electromagnetic coupling to the surroundings.

The second mechanism affecting the delay losses is the quality factor of the varactors. The quality factor of each varactor topologies are discussed in detail in the previous sections. When the lines are loaded, the overall insertion loss of the microstrip line topology is almost twice that of the loaded CPW topology. This is due to the fact that the quality factor of the varactor used in loaded CPW is more than 120 and that of used in loaded microstrip line is less than 18.7. For a general purpose, η_{VDL} can be improved if a low-loss transmission line can be loaded with varactors having high quality factors. Due to similar reason, the loaded slotline provides a better η_{VDL} than the loaded microstrip line even though the slotline itself has the highest insertion loss.

Compactness

Another point to be discussed is the compactness of the delay lines. The compactness refer the minimum ℓ_{physical} required for providing 360° differential phase shift at

[3] Here, the normalization factor is chosen to be $\ell_{\mathrm{physical}}/\lambda_0$ instead of ℓ_{physical} to exclude the effects of the operating frequencies on the physical length.

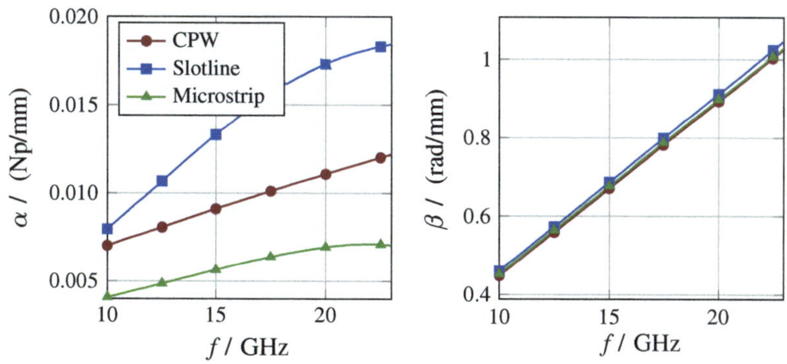

Fig. 4.28 *Left* The attenuation constants α of the not tunable transmission lines of the loaded variable delay lines. *Right* The propagation constants β of those lines. The characteristic impedances are $Z_{\text{Line, CPW}} = 100\,\Omega$, $Z_{\text{Line, Slotline}} = 75\,\Omega$ and $Z_{\text{Line, Microstrip}} = 110\,\Omega$

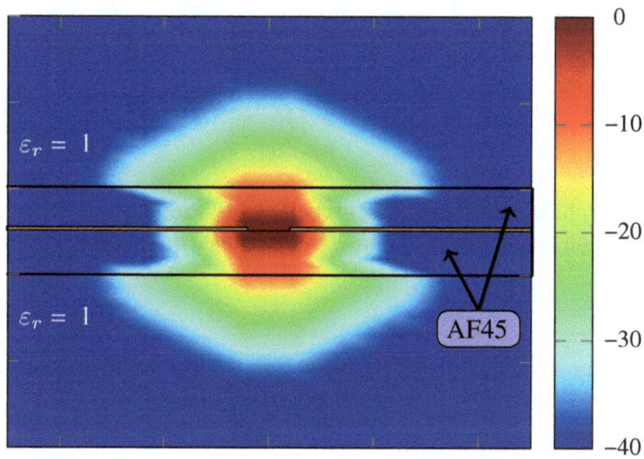

Fig. 4.29 Cross-sectional view of the slotline, where the amplitude of the E-field is plotted at 18 GHz. The colorbar is given in dB scale

their operating frequencies. The normalized ℓ_{physical} of the delay lines can be seen in Table 4.8. The loaded line topologies are 2 to 4 times shorter than the IMSL since they utilize the slow-wave effect [Poz11, HY71]. This can be seen in Fig. 4.30. There, phase of the transmission coefficients are plotted for two cases of each proposed structures. The first one is the phase of the *loaded* delay line and the second one is phase of a *not loaded* structure, which has the same physical length as the loaded one. The loaded lines are electrically much longer than the not loaded ones due to the capacitive loading and so, their phase responses are steeper. The ratio between the two curves at each frequency point can be defined as a miniaturization factor, which shows the relative compactness of the structures. The miniaturization factors are 2.18, 2.38 and 5.54 for the loaded CPW, slotline and microstrip line, respectively, at their operating frequencies.

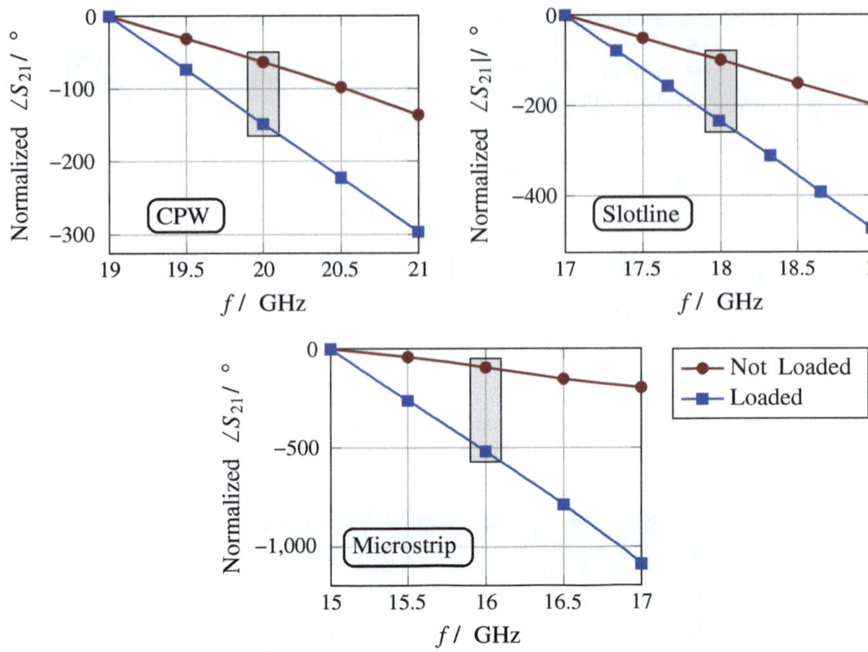

Fig. 4.30 Normalized phase of the transmission coefficients. Proposed delay lines are compared to their not loaded lines with the same physical length. The *shaded areas* show the operating frequency of the devices

In a phased array application, the delay lines have to be implemented into the limited area, which is defined by the element spacing of the array. Typically, this area is less than $0.65\lambda_0 \times 0.65\lambda_0$ for a 2D beam steering arrays. The loaded CPW delay line is the shortest device with a physical length of $1.773\lambda_0$. The delay line has to be meandered to fit the structure into the limited area. One challenge is to prevent the electromagnetic coupling between the electrodes. For this purpose, full-wave simulations are required. Another issue is to keep the desired transmission line operation mode. In Fig. 4.31, phases of the surface current densities are given for the three transmission lines. The structures include two 90° bends and they are simulated by using CST Microwave Studio. For the CPW line, the physical length of the outer ground electrode is higher than that of the inner ground to form the curve. As can be seen in Fig. 4.31a, the signal on the outer electrode delays compared to the one on the inner ground plane after the CPW is bended. As a result, the phase front (dashed lines in Fig. 4.31) is not orthogonal to the transmission line. After the second bending, the delay increases more and the desired CPW mode is disturbed. Since then, the line impedance changes and the delay line functionality disappears. The desired operation mode can be kept by using air bridges at the bending points to equalize the signal potential and the phase [Sim01]. Nevertheless, the fabrication effort and the costs increase considerably. All the advantages of the loaded CPW

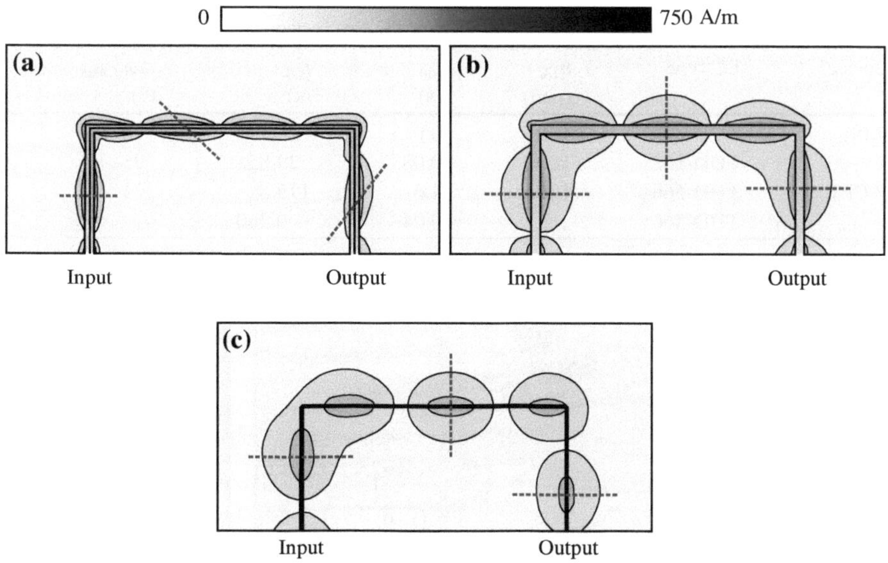

Fig. 4.31 Phase of the surface current density for the bended. **a** CPW, **b** slotline and **c** microstrip line at their own operating frequencies

delay line are available if the structure can be bended without disturbing the CPW mode. Otherwise, the loaded slotline or microstrip delay lines can be utilized because this problem is not encountered as can be seen in Fig. 4.31b, c, respectively.

Response Times

The beam steering rate of a phased arrays is determined by the response times of the variable delay lines, which are t_{on} and t_{off} (see Sect. 2.2.2). The main factors effecting the response times are the height of the LC layer h_{LC}, the LC material itself and the tuning mechanism. In the frame of this work, several IMSL devices with different alignment layer parameters are fabricated and tested to improve the functionality of the alignment layer. These results are presented in Appendix A.1.2. In fact, significant improvements, which reduces the response times, are achieved by using the loaded line topologies since they enable to use a low h_{LC} [Goe09]. In Table 4.9, measured response times of different devices are listed. The results confirm the fact that response times of LC devices are proportional to the square of h_{LC}. Additionally, it is shown that the response times of the IMSL can be reduced from 175.8 s to 8.13 s by using an LC material with a less rotational viscosity (see Sect. 2.2.2). The drawback is the reduced tunability of the material and therefore, $\Delta\phi_{\beta,\text{max}}$ is decreased at the same time.

The response times are given for two extreme alignments of the director, which are either from perpendicular to parallel or parallel to perpendicular alignments with

Table 4.9 Measured response times and $\Delta\phi_{\beta,\max}$ of the *LC* devices

Device	LC type	h_{LC} (μm)	t_{on} (s)	t_{off} (s)	$\Delta\phi_{\beta,\max}$ (°)
IMSL	TUD-823	100	0.013	8.13	76
	TUD-862	100	0.083	29.85	95
V-CPW	TUD-566	100	4.6	175.8	274
	TUD-566	5	0.04	0.260	–

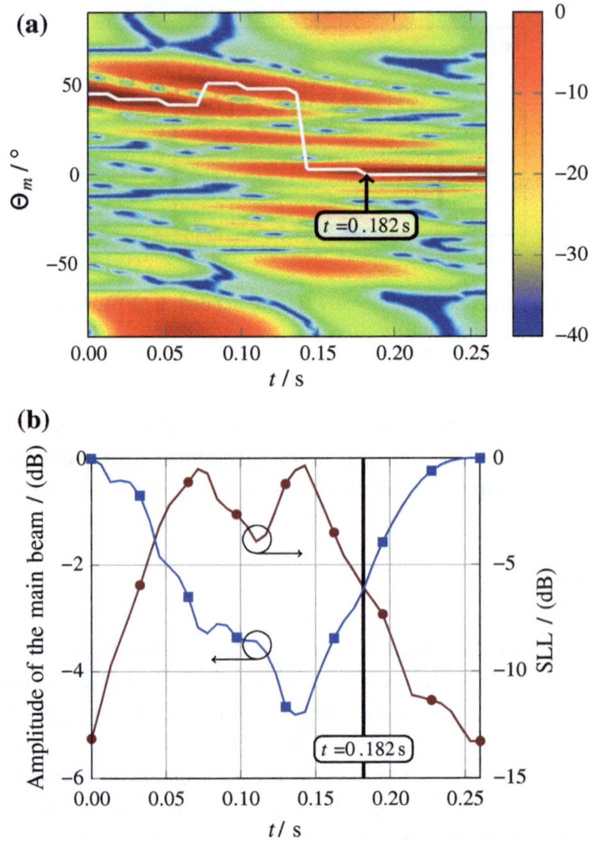

Fig. 4.32 **a** 2D plot of the magnitude of the antenna pattern versus the time t. The *white line* highlights the beam direction Θ_m at each time. The colorbar is given in dB scale. **b** The amplitude of the main beam and the corresponding side lobe level (SLL) versus the time t

respect to the electric field of the IMSL. In an antenna application, on the other hand, the LC molecules are tuned in between these two cases instead of a complete cycle [Fri+11]. Thus, the time required to obtain a desired differential phase shift is less than the response time of the delay line. An example is given in Fig. 4.32, where the array factor AF is plotted over the time t. The antenna array includes 16×16 elements with $0.55\lambda_0$ element spacing. The delay lines are assumed to be a loaded CPW line, which provides $\Delta\phi_{\beta,\max} = 360°$ in $t_{\text{on}} = 0.04$ s and in $t_{\text{off}} = 0.26$ s, accordingly. For the given scenario, the antenna main beam direction Θ_m points out to 45° initially. This can be accomplished by applying bias voltages to the delay lines. As being a worst case scenario, all the bias voltages are switched off. Thus, the LC molecules are aligned by the alignment layer with the highest response time being effective for all delay lines. In this case, the main beam is steered to the broadside direction ($\Theta_m = 0°$). A MatLab script is written, which calculates the initial differential phase distributions of the delay lines, leading to $\Theta_m = 45°$. The final differential phase distribution is a zero matrix since $\Theta_m = 0°$ at the end. Any other phase distributions between these two states are calculated by assuming that the differential phase shift is linearly proportional to the time. Figure 4.32a shows a 2D plot of the magnitude of the array factor (AF) over the time. The main beam direction versus the time is plotted on the same figure as well. Although $t_{\text{off}} = 0.26$ s, the main beam directs to 0° after $t = 0.182$ s. At this point, many of the delay lines reach to their desired phase values but not all. This is enough to point towards $\Theta_m = 0°$, whereas the side lobe levels (SLLs) are still high. In Fig. 4.32b, the magnitude of the main beam and the SLL are plotted versus the time. When $t = 0.182$ s, the amplitude of the main beam is 3 dB less than its maximum and the SLL is −6 dB. The optimum array factor with the highest magnitude and the lowest SLL is obtained when all delay lines reach their final states. The time required for the optimum AF is almost equal to t_{off}. Similar results are presented in details in [Fri+11].

Reliability

In LC devices, the critical fabrication parameter is the height of the LC layer h_{LC} since the electrical properties of the devices are defined by the LC material if the fringing fields are neglected. In this work, a desired h_{LC} is accomplished by using either glass made micropearls or a negative photoresist SU-8. The micropearls are preferred when a high h_{LC} is needed, e.g. 100 μm or more, and the usage of SU-8 is convenient to obtain a low h_{LC}, i.e. a few μm. Two delay lines are chosen to show the impact of uncertainty in h_{LC} on the device performances: IMSL and the loaded slotline. The first one is the most frequently used LC delay line in the literature. The second one is chosen because its varactor is implemented in parallel plate topology, which has high sensitivity to h_{LC}.

According to the confocal microscope measurements, the fabrication tolerances in h_{LC} are in a range of ±10 %. Nevertheless, for the investigations, it is assumed that the tolerances are around ±20 % at least. Then, the full-wave simulations are performed for a given h_{LC} and for the two relative permittivities of the LC layer, which are $\varepsilon_{r,\perp}$ and $\varepsilon_{r,\parallel}$. The results are depicted in Fig. 4.33. The maximum differential phase shift and the characteristic impedance of the IMSL are affected in a opposite manner by the h_{LC}. If h_{LC} is less than the optimum value, which is 100 μm in this case, the capacitance of the line increases. As a result, its tunability, and therefore,

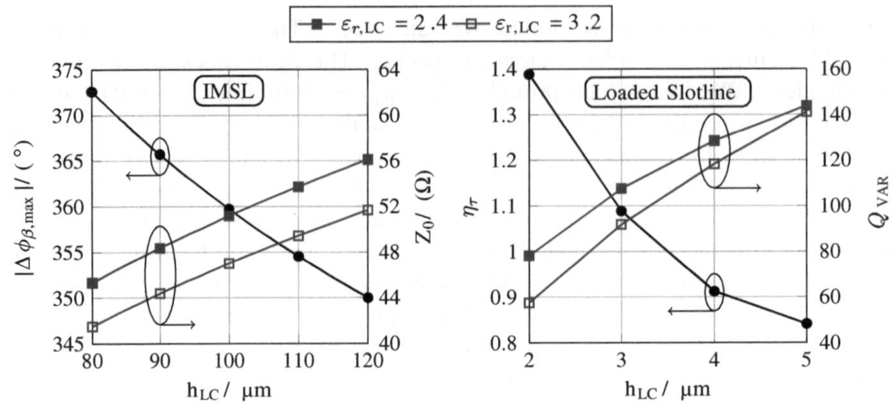

Fig. 4.33 Effects of the liquid crystal height on delay line performances. *Left* The maximum differential phase shift and the characteristic impedance of the IMSL at 20 GHz. *Right* The tuning efficiency η_T and quality factor Q_{VAR} of the varactors used in the loaded slotline at 18 GHz

$\Delta\phi_{\beta,\max}$ increases and the characteristic impedance Z_{Line} decreases. It should be noted that both, $\Delta\phi_{\beta,\max}$ and Z_{Line} change slightly, which are about $\mp 2.7\,\%$ and $\pm 10\,\%$, respectively.

Similar results are obtained for the parallel plate varactor used in the loaded slotline. The tuning efficiency of the varactor η_T and its quality factor Q_{VAR} are affected more compared to IMSL's performances. For instance, if h_{LC} deviates $\pm 1\,\mu$m from its optimum value, which is supposed to be 3 μm, Q_{VAR} changes from 60 to 120. These results are also applicable to the other loaded line devices.

Low response time and compactness are the main advantages of using a loaded line topology whereas all the drawbacks mentioned herein have negative impacts on the device performances. For instance, they have high insertion losses, resulting low efficiencies. On the other hand, the IMSL topology offers a high performance. Its disadvantage is high response time, which leads to a slow beam steering. Indeed, the beam steering rate is defined by a targeted application. For instance, low beam steering rate is accepted for a *portable* applications such as satellite TV reception in caravans since the satellite tracking is done once when the terminal is stationary. In contrary, low response times and so, a fast beam steering is necessary for *mobile* applications such as satellite TV reception on-the-move. In both cases, the antenna needs low loss delay lines, which can be accomplished by using IMSLs. Additionally, the response times of the IMSLs can be reduced by a factor of 20 or even up to 40 times by using an LC material with a low viscosity and by reducing h_{LC} from 100 μm to 50 μm, simultaneously. Even in this case, its efficiency will be around 60°/dB. In the lights of these considerations, the IMSL type delay line is chosen for a 2D beam steering phased array, which is subject to the next section.

4.3 Electronic Beam Steering Phased Array

This section presents, the design and measurement results of a novel 2D electronic beam steering phased array. A multilayer antenna topology is proposed for a flat and low profile phased array. A demonstrator, which is designed to operate at 17.5 GHz, includes a 2 × 2 microstrip patch antenna array, LC based variable delay lines, an RF-feeding network, and a dc-biasing network. This is a small prototype for a real application, however enough to verify the proof-of-concept. In this section, first, a description of the antenna array is given to clarify the topology and to point out the links between the different components. Second, the design and measurement results of the individual components are presented. Third, the scattering parameters and the far-field measurement results of the antenna demonstrator are discussed. Finally, the performances of a large array of the proposed demonstrator are addressed.

4.3.1 Antenna Concept

Schematics of the phased array demonstrator are depicted in Fig. 4.34a–b from the top and bottom views, respectively. As shown in the cross-sectional view in Fig. 4.34c, the microstrip patch antenna array [BP80] is fabricated on the top side of the front substrate. The ground electrode is implemented on the back side of the front substrate. Due to that, the electrical characteristics of the microstrip patches such as the efficiency and the bandwidth are determined by the front substrate. 700 μm thick BF33 type glass is chosen to improve the both characteristics since it has the lowest relative permittivity with the largest thickness [Bal05]. The ground electrode includes rectangular coupling slots overlying and centring the microstrip patches to form an aperture coupling (see [Poz85]) between the patch antennas and the IMSL variable delay lines. The signal electrodes of the IMSLs are mounted on the top side of the back glass substrate. The tunable LC material (TUD-566) is contained between the two glass substrates, where the top substrate is elevated 100 μm by means of micropearls. These electrodes are coated by the alignment layer to anchor the LC molecule parallel to the surface. Contrary to [KFN02], the delay lines are meandered in spiral geometry as shown in Fig. 4.34b. This geometry features several improvements, which are discussed in the subsequent section. The variable delay lines are connected to the RF-feeding network through a dc-blocking structure to bias the delay lines separately. In principle, the dc-blocking structure delivers the RF signal and blocks the bias voltage in order not to affect the other delay lines. The RF-feeding network is implemented in IMSL topology as well. Nevertheless, its electrical parameters are not tuned because of the dc-blocking structure.

The antenna demonstrator can be used either in receiving or transmitting mode since it is a passive and reciprocal device. Nevertheless, the operating principle is given only for a receiving mode to explain the device in a clear way. In this case, the RF signal received by the radiating element is coupled to the IMSL via the aperture

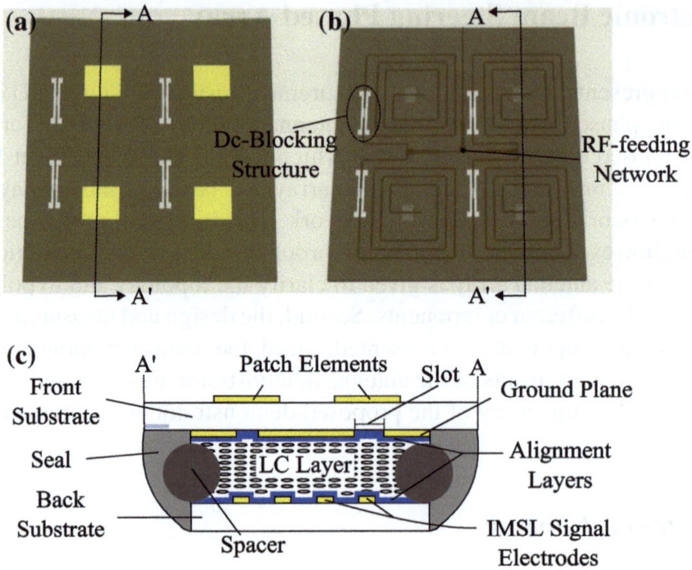

Fig. 4.34 Schematics of the phased array based on LC display technology. **a** Top, **b** bottom and **c** cross-sectional views

coupling. The signal propagates through the IMSL towards the RF-feeding network. At this time, the dielectric properties of the LC material and therefore, the phase of the RF signal is tuned by applying a bias voltage across the signal and ground electrodes of the IMSL. The signal is then electromagnetically coupled to a CPW, which is implemented on the IMSL's ground electrode. After propagating along the short CPW line, the RF signal is coupled to the RF-feeding network. By this way, a contactless RF interconnection is achieved between the IMSL and the RF-feeding network, operating as a dc-blocking structure. The LC material is completely filled between the front and back substrate. It is tuned only underneath the IMSL's signal electrode since the biasing voltage can not reach to the other components due to the dc-blocking structure. The bias voltages are delivered to the IMSLs through resistive bias lines. These lines are made of Chromium (Cr).

The RF-feeding network combines the identical unit elements. The unit element consists of an aperture coupled microstrip patch antenna, an IMSL and a dc-blocking structure. The size of the unit element is limited, which is defined by the element spacing of the microstrip antennas. In fact, the area should be even less than the element spacing because the RF-feeding network has to be implemented between the unit elements as well. All these components are designed and verified before a complete array is built. In the following section, these results are presented in detail.

4.3.2 Design and Verification

Aperture Coupled Microstrip Patch Antenna Element

A microstrip patch antenna given in Fig. 4.35 is fabricated and filled with LC. Its dimensions are given in Appendix A.4.4. It is aimed to measure the performance of the aperture coupling and the resonance frequency of the patch antenna. The structure also includes a transition from CPW to microstrip line for the on-wafer characterization.

The patch dimensions are first calculated with the formulas given in [Bal05]. Later on, they are optimized together with the slot dimensions by using ADS Momentum. The simulated reflection coefficients presented in Fig. 4.36 show that the antenna operates at 17.9 GHz with a 6.1 % 10 dB fractional bandwidth. Its gain and radiation efficiency are determined as 4.44 dB and 0.60 at the operating frequency.

In Fig. 4.36, the measured reflection coefficients are presented as well. The antenna operates at 18.35 GHz with a 6.5 % bandwidth. The resonance frequency of the fabricated structure is slightly higher that is about 2.5 % than the simulated results. During the simulations, the electrical parameters of BF33 were used as they were defined by the manufacturer. The shift of the resonance frequency to the higher frequency indicates that the actual relative permittivity of the BF33 is lower than the one used in the simulations. This confirms the results obtained in Sect. 3.2.

DC-Blocking Structure

DC-blocking structures are placed between the delay lines and the RF-feeding network to bias the delay lines independently for a 2D beam steering. Its operating

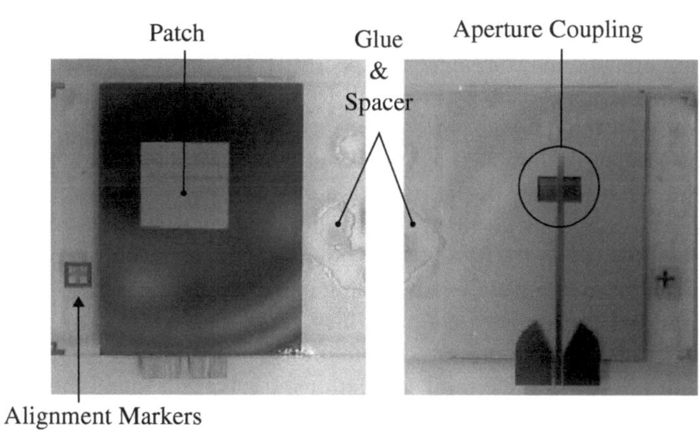

Fig. 4.35 **a** Top and **b** bottom view photographs of an aperture coupled microstrip patch antenna element. The patch size is 3.4 mm × 3 mm and slot size is 1.5 mm × 1 mm

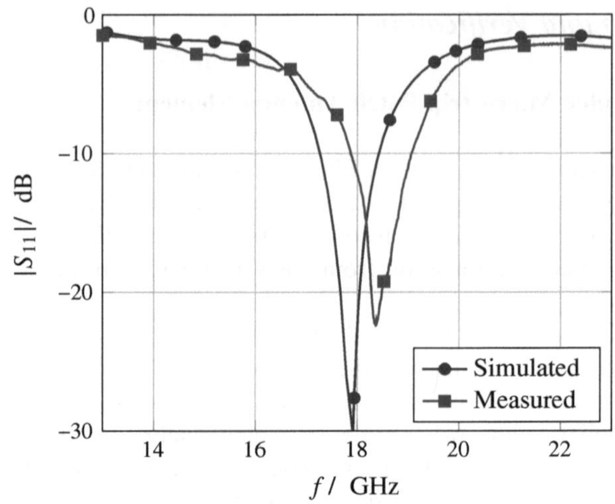

Fig. 4.36 Magnitude of the measured and simulated reflection coefficients of the aperture coupled microstrip patch antenna element

principle is the electromagnetic coupling of the RF signal between the microstrip line and CPW, where the latter is located on the ground electrode of the former. The coupling performance of the transmission lines is significant because all the energy radiated by one line has to be coupled to the other. Otherwise, an undesired radiation occurs and the antenna efficiency decreases.

Photographs and the dimensions of the structure are given in Fig. 4.37 and in Appendix A.4.4, respectively. Figure 4.38 shows the simulated and measured scattering parameters of the dc-blocking structure. When the structure is matched ($|S_{11}| < -10$ dB), the transmission is maximized. This confirms that the RF signal propagates along the structure and there is no remarkable undesired radiation. Specifically, the insertion losses are determined as 1.23 dB and 1.5 dB at 17.5 GHz for the simulation and measurements, respectively. It should be noted that these values include the losses caused by the CPW to microstrip line transitions, which are necessary only for the on-wafer measurements. The corresponding measured return loss is 23 dB, which is simulated as 18.5 dB at the operating frequency.

The frequency response of the transition can be further optimized by choosing proper layout dimensions so that, it transmits/receives the signal only at desired frequencies similar to an RF filter. In the antenna application, this can be utilized to suppress the undesired signals without using an additional filter.

An interdigital capacitor or coupled lines could also be used as a dc-blocking structure. Nevertheless, the proposed structure is chosen after a comparison of different possibilities by doing circuit simulations. It has been found that for a similar RF performance, the required electrode dimensions and the gaps between the electrodes are relatively large for the proposed structure. This reduces the fabrication

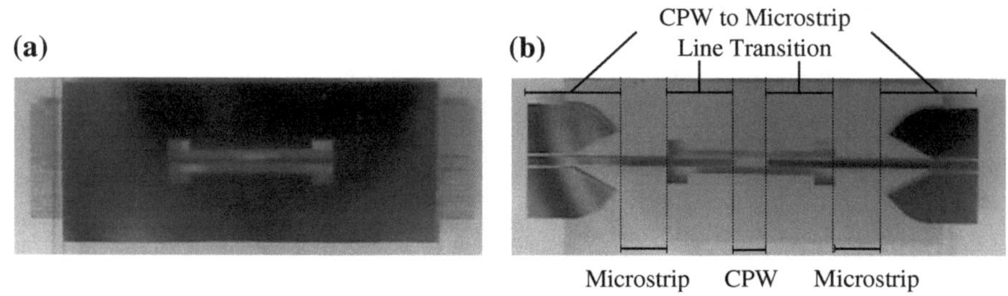

Fig. 4.37 **a** Top and **b** bottom view photographs of the DC-blocking structure

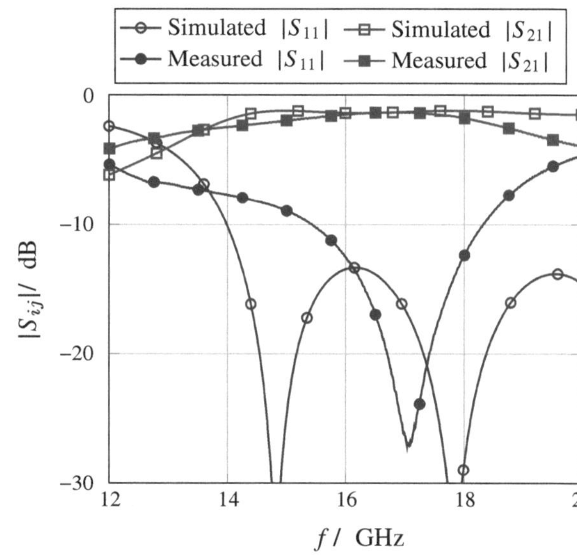

Fig. 4.38 Magnitude of the measured and simulated scattering parameters of the dc-blocking structure

effort and increases the yield of the antenna since the dc-blocking structure is needed for all unit elements.

Variable Delay Line in Spiral Geometry

In [KFN02] and similarly in Sect. 4.2.1, the variable delay lines are meandered to obtain a high $\Delta\phi_{\beta,\max}$ in a compact way. Up to now, these devices have been considered only as single tunable components. However, several issues arise when their integration is aimed into an antenna system. Firstly, the input port of the IMSL, where the aperture coupling takes place, has to be located under the center of the microstrip patch antenna. Due to that, the delay lines have to be shifted physically as shown in Fig. 4.39 with respect to the radiating elements. This is a minor problem, which results in losing the design symmetry. Secondly, the RF-feeding network requires

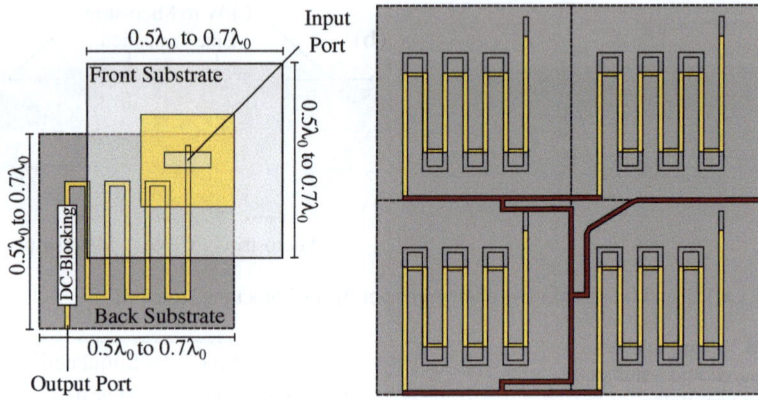

Fig. 4.39 Difficulties when a meandered IMSL is integrated into an antenna array. *Left* A unit element showing that the back substrate is shifted with respect to the front substrates. *Right* Area consumption and complex RF-feeding network is required to combine the delay lines. The delay lines are shortened to save space for the RF-feeding network

relatively more area. This is due to the fact that the output port of the IMSL, where the IMSL is connected to the RF-feeding network, are not close to each other as illustrated in Fig. 4.39. As a result, the RF-feeding network consists of long transmission lines and several bendings. Additionally, the delay lines would be shorter because of the area consumption of complex RF-feeding network.

In this work, a spiral geometry is proposed first time for variable delay lines to overcome these issues. The IMSLs are implemented in a spiral geometry as shown in Fig. 4.40. The coupling of the RF signal between the delay line and the radiating element is accomplished in the middle of the structure. As an advantage, the front and back substrates are centered, enabling design symmetry to be preserved. Additionally, the geometry allows flipping the delay line along a horizontal symmetry axis passing through the center. Hence, the input port of the delay line shifts to the other side, whereas the coupling point still remains in the center. For the antenna, this leads to design a more compact RF-feeding network. Simultaneously, the distance between the radiating elements is kept constant, which is a fundamental requirement for uniform arrays.

The characterization of a spiral delay line is not feasible as a single component, since its input port is located inside the LC sandwich structure. In order to perform the measurements, low-loss RF vertical interconnects are required through the glass substrate. This is not a standard technique used in LC display technology, and therefore, it is out of the scope of this work. Nevertheless, the spiral and meander delay lines are compared by means of full-wave simulations using CST Microwave Studio. The layout of the spiral delay line is given in Appendix A.4.4. The results are depicted in Fig. 4.41. Both delay lines are designed on the same area by using identical design rules, i.e. minimum gap $g_{\min}$ between the two electrodes are identical. The overall length of the spiral delay line is 5 mm longer than the length of the meander line,

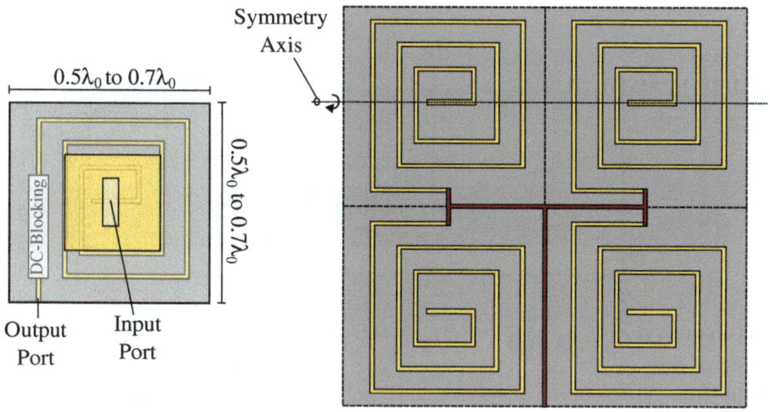

Fig. 4.40 Advantages of implementing the IMSL in spiral geometry. *Left* A unit element showing that the front and back substrates are centered. *Right* Compact and simple RF-feeding network is able to combine the delay lines

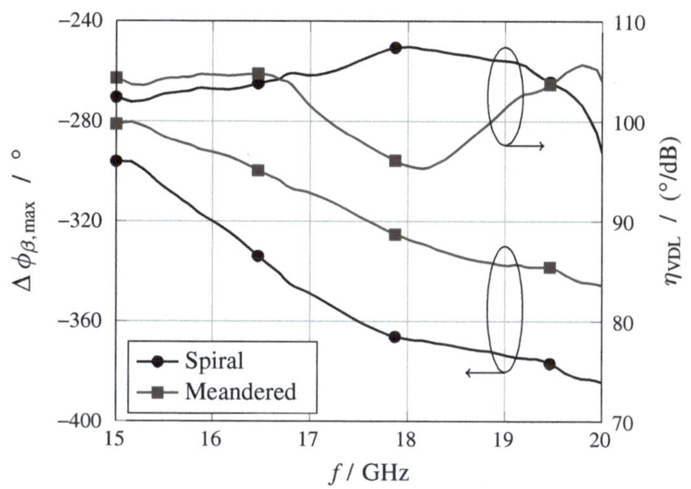

Fig. 4.41 Comparison of the maximum differential phase shift $\Delta\phi_{\beta,\max}$ and the efficiencies η_{VDL} of the spiral and meandered delay lines. The delay lines are simulated without the CPW to microstrip line transitions

which is 75 mm. This is due to the fact that the limited area is used more efficiently, being another advantage of the spiral geometry. Hence, $\Delta\phi_{\beta,\max}$ of the spiral delay line is increased.

Bending a transmission line increases its fringing fields since its capacitance and therefore, the characteristic impedance deviates from the desired value [Gup+96]. This leads to an undesired coupling to the surrounding components. Figure 4.42 shows the normalized $\vec{E}$ field amplitude distribution of a bended electrode, which are obtained by using CST Microwave Studio. The structure includes different types

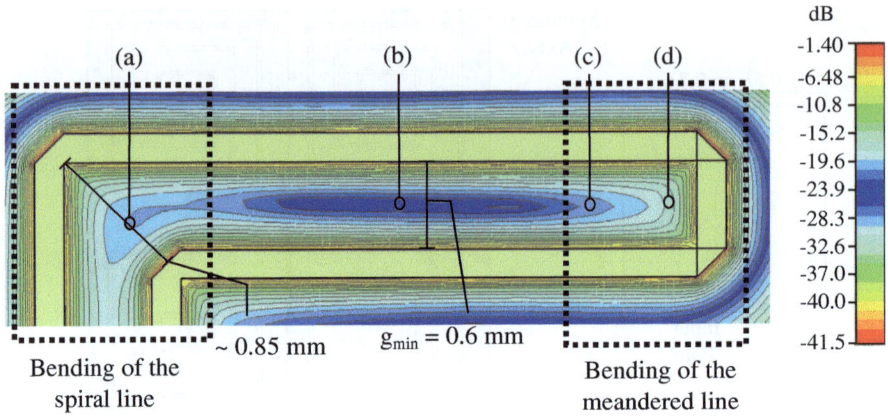

Fig. 4.42 Amplitude of the E-Field distribution for two types of bendings, which model the spiral and meandered lines. The maximum field strength is normalized to 0 dB by CST Microwave Studio. The field strengths are **a** and **c** −20 dB, **b** −28 dB and **d** −12 dB at 20 GHz

of bends, modelling those utilized in the spiral and meandered delay lines. The gap between two parallel electrodes $g_{\min}$ is specified to be 0.6 mm, resulting a negligible field strength that is less than −28 dB. When the electrodes are bended, the field strength increases up to −12 dB if the electrode is bended like in the meandered line. In contrary, this value can be still kept less than −20 dB by using the spiral geometry because the distance between two adjacent bends is equal to $\sqrt{2} \times g_{\min}$. Therefore, although the spiral delay line is 5 mm longer than the meandered line, its insertion losses are almost kept constant since the slight couplings at the bends are avoided. The resultant η_{VDL} of the spiral delay line is increased, which is expected to be in the range of 105°/dB at the operating frequency. In the antenna demonstrator, the spiral IMSLs are utilized because of the presented improvements.

RF-Feeding and Biasing Networks

The RF signal is distributed to all delay lines equally in amplitude and in phase through microstrip line power dividers. The impedance matching of the RF-feeding network is accomplished by using standard $\lambda/4$ impedance transformers. Figure 4.43 shows a overview of the back substrate. Although the structure is completely filled with LC, the electrical characteristics of the impedance transformers do not change when a bias voltage is applied. This is due to the fact that LC underlying the RF-feeding network is not tuned because of the dc-blocking structure.

The applied bias voltages are delivered to the RF circuitry through high resistive bias lines. The bias lines are implemented on the adhesive chromium (Cr) layer, which has a sheet resistance of 25.3 Ω/sq. The line width is set to be 10 μm to increase the bias line resistance. The measured dc resistances between the bias pads

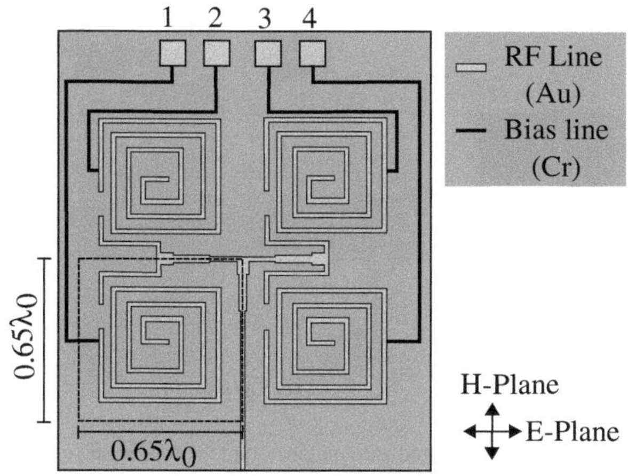

Fig. 4.43 Layout of the 2 × 2 antenna array prototype. The figure shows the components mounted on the top side of the back substrate. These components are RF-feeding network, variable delay lines, biasing network and four bias pads

and the corresponding delay lines are in the range of 160–900 kΩ, depending on the line length.

4.3.3 Measurement Results

Two antennas are realized and characterized as proof-of-concept. The first antenna is a steerable antenna, which is shown in Fig. 4.44. The second antenna is almost similar to the first one but it includes no bias lines and thus, it is not steerable. This is used as a reference antenna for quantizing the effects of the Cr bias lines on the antenna characteristics. Both planar antennas have an overall thickness of only 1.5 mm. They are installed with SMA and a four-pin dc connector for the measurements, which increases the overall profile of the antenna. In a real application, on the other hand, the received RF signal can be first down converted and then can be delivered to the output port by using a compact MCX or SMD RF connector, for instance I-PEX MHF connector, to keep the low profile.

The measured reflection coefficients of the prototype are presented in Fig. 4.45. For the zero biasing case ($V_b = 0\,\text{V}$), the main beam points out the broadside direction with a return loss better than 20 dB at 17.5 GHz. This value reduces to 15 dB when the antenna is steered. The operating frequency of the antenna is hardly affected by the applied voltage since the microstrip patch antennas are fabricated on the glass layer. Nevertheless, a slight change in the antenna bandwidth occurs, because the properties of the LC material has an impact on the aperture coupling efficiency. The reference antenna, on the other hand, has 10 dB better return loss than that of the steerable

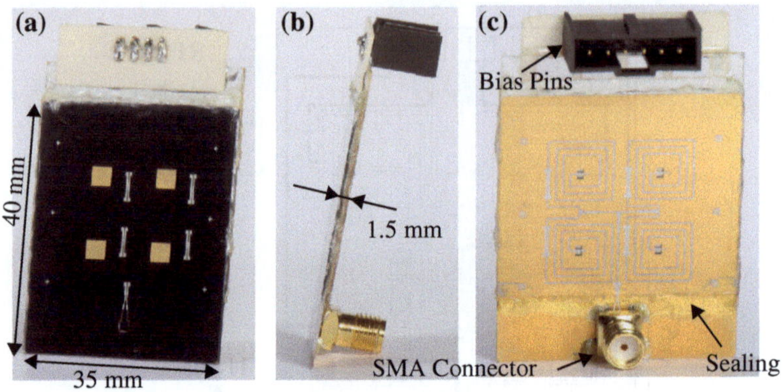

Fig. 4.44 Photographs of the fabricated antenna prototype. **a** Top, **b** side and **c** back views

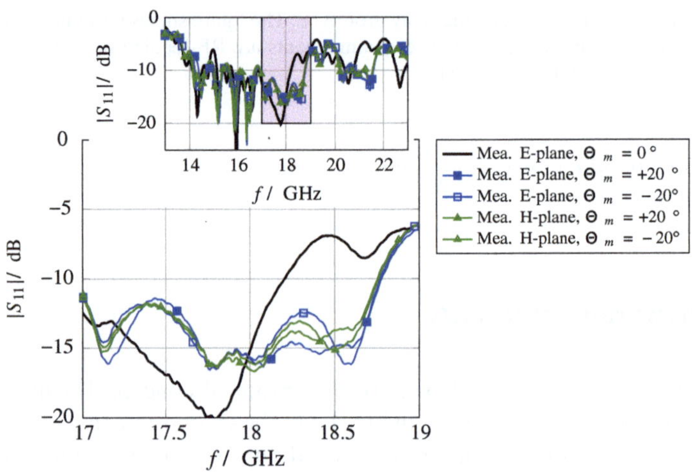

Fig. 4.45 Measured and simulated reflection coefficients of the phased array at different main beam directions Θ_m

antenna. Empirical studies show that the impact of the bias lines on the RF circuitry can be negligible if the line resistance is higher than 3 MΩ for the given structure. Although the Cr material affects the RF circuitry, the measured return losses are still practical, and therefore, considering the reduced fabrication effort, the bias lines can be implemented by using Cr adhesive layer.

The radiation pattern measurements of the phased array are performed in an anechoic chamber. The RF power received by the antenna prototype is recorded by a power meter. The antenna is mounted on top of a turn table and the beam steering in the E- and H-planes is measured separately to confirm the 2D beam steering capability.

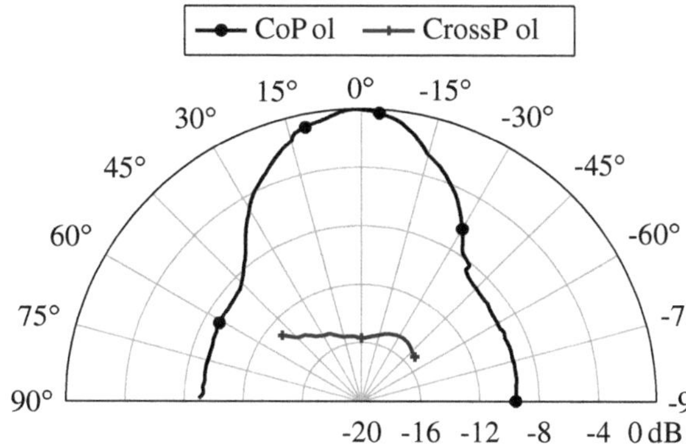

Fig. 4.46 Measured received power of the antenna prototype for co- and cross-polarizations

The antenna prototype receives linearly polarized signals. In Fig. 4.46, the received powers are plotted for co- and cross-polarized incoming signals. As it is shown, the cross polarization level of the antenna is a range of 13–16 dB. The microstrip antennas can be designed to operate in a circular polarization by changing either their feeding point or geometry [BP80].

The variable delay lines in the same column are biased together to steer the main beam since the antenna includes only 2×2 elements. Only two bias voltages, one for each column, are applied through a bias tee. In fact, one of these voltages is varied while the other one is kept at 0 V. The antenna gain G is determined for different main beam directions Θ_m by using the two antenna method [Bal05]. The measured and estimated antenna gains in E- and H-planes are plotted in Fig. 4.47 for different beam directions. The measured gains are 0.5–1.5 dB less than the estimated values because of two expected reasons. First, the Cr bias lines increase the delay line insertion loss, which reduces the transmission and therefore, the antenna gain. Second, the connections and adaptors used in the measurement setup add additional losses. Furthermore, the measured gains in the E-and H-plane match quite well. Slight differences that is less than 0.8 dB are caused by the fabrication tolerances.

Figure 4.48 shows a strong dependency of the antenna gain with the operating frequency. This is mainly caused by the poor matching of the aperture coupled microstrip antennas at the given frequencies. According to Fig. 4.36, the matching of the aperture coupled microstrip antenna is around -3 dB at 16 GHz and it is about -20 dB at 18.3 GHz. Therefore, the overall efficiency and the gain of the prototype reduce considerably at lower frequencies ($f < 17$ GHz). This can be overcome by redesigning of the radiating elements or choosing another type of radiating elements, which provide more bandwidth.

The measured far-field radiation patterns are given for in the E-and H-planes in Figs. 4.49 and 4.50, respectively. The applied bias voltages and the corresponding main beam directions listed in Table 4.10. The measurement results obtained in the E-plane are compared to the simulation results, which shows a good agreement of the simulated and measured data for a given beam direction.

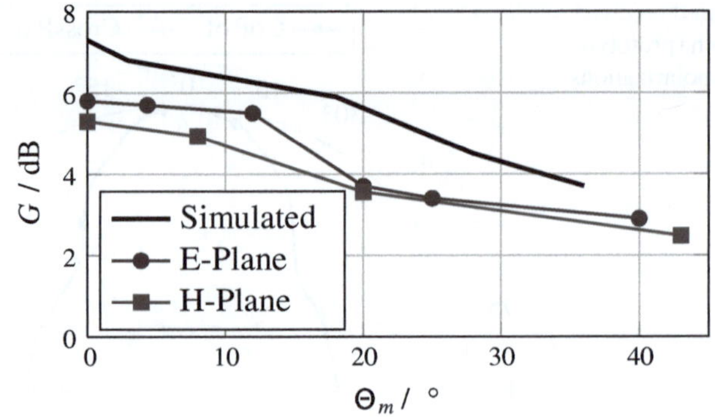

Fig. 4.47 Comparison of the estimated and measured antenna gain G for different main beam directions Θ_m at 17.5 GHz

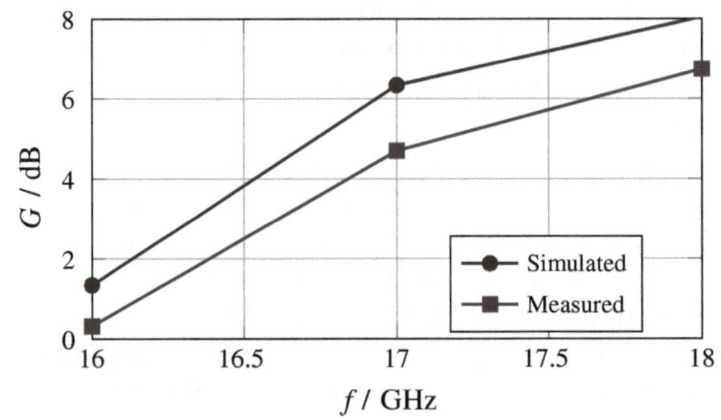

Fig. 4.48 Deviation in the antenna gain G versus the operating frequency at the broadside ($\Theta_m = 0°$)

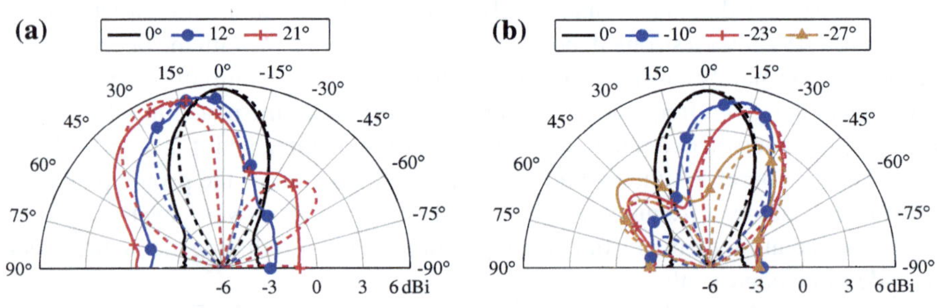

Fig. 4.49 Measured and simulated antenna gains along the E-plane for **a** $\Theta_m > 0°$ and **b** $\Theta_m < 0°$. The *solid* and *dashed lines* represent the measured and simulated values, respectively. The measured raw data are smoothed by using MatLab 5th order moving average function. The *dashed lines* are normalized to the maximums of the corresponding measurements

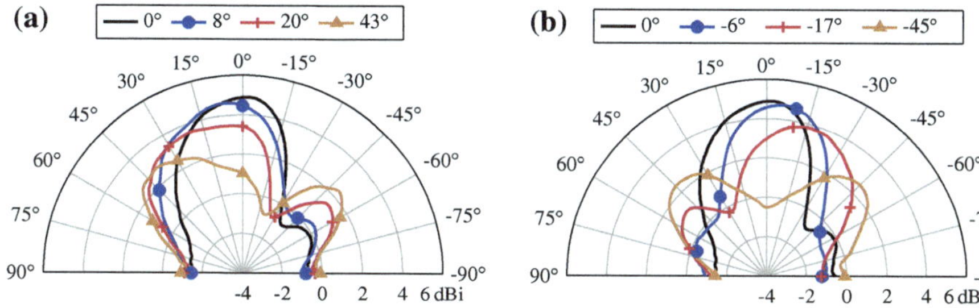

Fig. 4.50 Measured antenna gains along the H-plane for **a** $\Theta_m > 0°$ and **b** $\Theta_m < 0°$. The measured raw data are smoothed by using MatLab 5th order moving average function

Table 4.10 Applied bias voltages for the beam steering

Bias voltages				Beam steering	
$V_{b,1}$ (V)	$V_{b,2}$ (V)	$V_{b,3}$ (V)	$V_{b,4}$ (V)	E-plane (°)	H-plane (°)
0	0	0	0	0	0
0	8	8	0	12	0
0	11	11	0	21	0
7	0	0	7	−10	0
12	0	0	12	−23	0
13	0	0	13	−27	0
0	0	9	9	0	8
0	0	10	10	0	20
0	0	15	15	0	45
8	8	0	0	0	−6
10	10	0	0	0	−17
15	15	0	0	0	−45

The antenna gain reduces and side lobe level (SLL) arises at wide-scanning angles, because of being a small demonstrator. The antenna includes only 2 × 2 radiating elements with an element spacing of $0.65\lambda_0$, which is indeed too large. For instance, for $\Theta_m = \pm 25°$, the gain loss is about 2 dB and the SLL is −4 dB, which are also confirmed by the simulations. In particular, as shown in Fig. 4.50, when the main beam is steered to $\Theta_m = -45°$ a grating lobe at $\Theta_m = 45°$ occurs that is almost equal to the main lobe.

Beside the antenna gain and the far-field pattern measurements, the beam steering speed of the prototype is measured as well. In a first step, the polarization of the transmitting antenna and that of the antenna prototype (in receiving mode) are matched at the broadside. All the bias voltages are set to 0 V and so, the received power is maximized at $\Theta_m = 0°$. Then, the main beam direction of the prototype is steered to 45° by applying $V_{b,1} = V_{b,2} = 0$ V and $V_{b,3} = V_{b,4} = 15$ V. The change in the received power is recorded over the time. These results are plotted in Fig. 4.51. In this case, the received power reduces versus the time since the antenna prototype

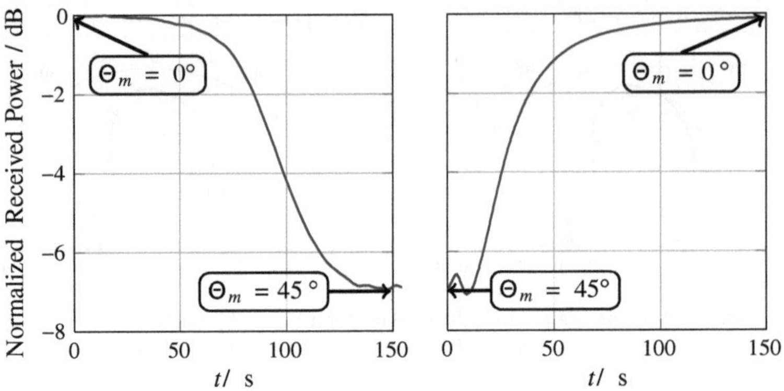

Fig. 4.51 *Left* Normalized received power over the time t when the main beam is tilted from 0° to 45° by applying bias voltages. *Right* The main beam direction is steered from 45° to 0° by the help of the alignment layer. In both cases, the transmitting antenna is located at $\Theta_m = 0°$

is steered to $\Theta_m = 45°$ gradually. In a second step, the bias voltages are again set back to 0 V and so, the LC molecules are oriented by the alignment layer. The change in the received power is again recorded and the result is plotted in Fig. 4.51. In contrary to the first step, the received power is maximized over the time because the antenna is steered back to the broadside, where the transmitting antenna is located. The measured beam steering speed that is about 150 s to tilt the main beam from $\Theta_m = 0°$ to $\Theta_m = 45°$ and vice versa. The empirical results presented in Sect. 4.2.5 is verified here since the measured beam steering rate is lower than the response time of the delay line, which is about 180 s. In the previous sections, it has been also presented that the response time of an LC delay line differs considerably whether the molecules are aligned by applying the bias voltage or by using the alignment layer. However, these results show that the beam steering rate is almost equal for the both tuning mechanisms. This due to the fact that the overdrive method [YW06] is not utilized while the antenna is steered. In other words, to increase the beam steering rate, first a higher voltage than a saturated voltage should be applied for a short period and then the actual one should be set. The saturation voltage refers a minimum voltage, which is able to align the LC molecules completely parallel to the bias voltage [Goe09]. For instance, the saturation voltage of the IMSL is around 25 V as presented in Sect. 4.2.1. However, the results given in Fig. 4.51 are obtained by applying $V_b = 15$ V. In the light of these measurement results, one can conclude that a biasing electronic, which enables the overdrive method similar to the ones used in LC displays, is also required for the LC based antennas to reveal the benefits of the alignment driven by the bias voltage.

Fig. 4.52 Photograph of the fabricated larger antenna prototype, including 8 × 8 radiating elements. The picture is taken form the back side, where the spiral shaped variable delay lines can be seen

Larger Arrays Based on the Proposed Antenna Topology

The gain of the prototype is measured as 5.9 dB while radiating in the broadside. As it is well-known from the array theory [Bal05], the antenna gain can be increased by increasing its aperture size, equivalently the element number. As shown in Fig. 4.52, the proposed antenna layout can be scaled for a larger aperture, including 8 × 8 radiating element with an aperture size of less than 12 cm × 12 cm. However, in a real application, two to four times larger apertures are required.

According to CST Microwave Studio simulation results given in Fig. 4.53, the gain can be increased to 25 dB when the element number is enlarged to 16 × 16. The corresponding aperture size is about 18 cm × 18 cm and 15 cm × 15 cm for element spacing of $d = 0.65\lambda_0$ and $d = 0.55\lambda_0$, respectively. To increase the gain further to 33 dB, the element number and so, the aperture size have to be doubled in both directions. The scan losses are less than 3 dB as long as the SLL is less than −10 dB. It should be noted that the antenna profile, which is 1.5 mm, always remains constant while increasing the aperture size.

Effect of the element spacing can be seen on the SLL, which also limits antenna scanning range because of appearing grating lobes. For instance, a SLL less than −10 dB can be achieved up to $\Theta_m = \pm 30°$ if $d = 0.65\lambda_0$. On the other hand, the same condition is kept up to $\Theta_m = \pm 55°$ if $d = 0.55\lambda_0$. The element spacing of the prototype depends on the overall length and compactness of the delay line. For this purpose, another LC material with a higher tunability can be used or IMSLs can be substituted to convenient loaded delay lines, which are more compact.

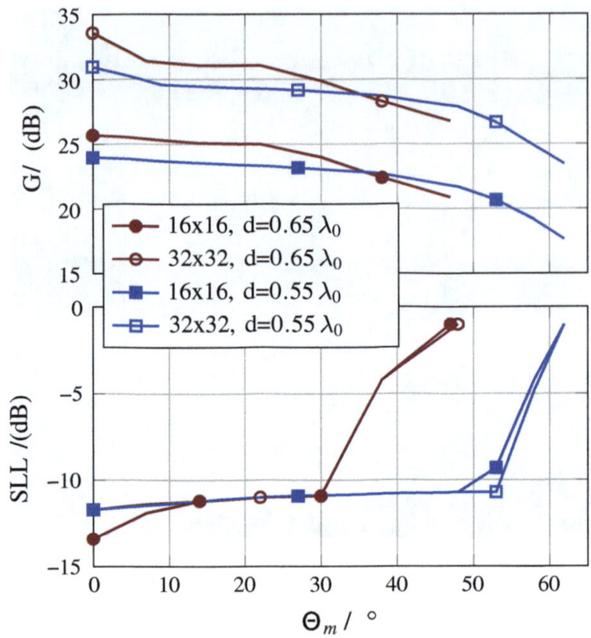

Fig. 4.53 Simulated antenna gain G and side lobe level SLL versus main beam direction Θ_m of large arrays of the proposed topology with different element spacing d and at 17.5 GHz

4.4 Summary

In this chapter, a 2D electronic beam steering planar phased array was presented in liquid crystal (LC) technology. The variable lines play an essential role concerning the performance, cost, and dimensions of the phased arrays. Thus, several LC based delay lines were investigated in a systematic way before the antenna was studied.

In the first section, the design methods were described by means of mathematical models for the transmission line and loaded delay lines. In the second section, four delay lines presented with their fabrication details and measurement results. These were inverted microstrip line (IMSL), loaded coplanar waveguide (CPW), loaded slotline and loaded microstrip line. The first two topologies had been studied by other researches as well whereas the two other topologies were studied for the first time in LC technology in this work. Particularly, the device layouts were optimized to utilize available material tunability efficiently. The results are listed in Table 4.11 and compared with other published data.

Considering the transmission line type delay line, high delay line efficiency was achieved in this work. This was accomplished by reducing the fringing fields by choosing layout dimension properly. Regarding to loaded delay lines, it was shown that the varactors' quality factor is essential to enhance the delay line efficiency η_{VDL}. For instance, the highest η_{VDL} was measured with the loaded CPW delay line

Table 4.11 Comparison of variable delay lines (VDL) at their operating frequencies f_0

Delay line Type	Line topology	References	Technology	f_0 (GHz)	η_{VDL} (°/dB)
Transmission line	Inverted microstrip line	[Kuk+02]	LC	20	18.5
		[BMS10]	LC	37	26
		[KFN02]	LC	10	30
		[Mue07]	LC	24	110
		This work	**LC**	**17.5**	**85** (Mea.) **105** (Sim.)
Loaded lines	Coplanar Waveguide (CPW)	[Bor+00]	RF MEMS	25	153.84
		[HR03]	RF MEMS	37.7	122.72
		[Jan+08]	RF MEMS	8.5	93.08
		[BR08]	RF MEMS	40	46.67
		[RSF05]	RF MEMS	9.0	39.50
		[NY99]	Semiconductor	20	85.71
		[PCS06]	Semiconductor	17.5	23.08
		[Goe09]	LC	20	60
		This work	**LC**	**20**	**62.1**
	Slotline	[CVH08]	RF MEMS	10	28
		[Mir+08]	BST	30	25
		[Muk+07]	BST	13	60
		This work	**LC**	**18**	**42.5**
	Microstrip line	[Saz+11a]	BST	10	38.05
		[Jan+08]	RF MEMS	8.5	50.72
		This work	**LC**	**16**	**25**

even though the CPW itself featured the highest attenuation. The line losses were compensated by means of low-loss varactors. The varactors' quality factor can be increased further by reducing the dielectric losses. As being a best case scenario, for example, a varactor can be implemented in metal-air-metal configuration if radio frequency microelectromechanical system (RF MEMS) technology is utilized. In this case, η_{VDL} is reported as 153°/dB at 25 GHz in [Bor+00] for a one bit delay line. The drawback is physically large varactors, resulting in longer or area occupying delay lines. This is again critical when a 2D beam steering antenna is aimed for, since the delay lines violates the compactness. Furthermore, the one bit resolution is not convenient for a beam steering antenna. This can be solved by implementing an additional biasing network, which might increase the insertion loss, depending on the material used for the bias lines.

Additionally, other issues such as the compactness, response time and reliability of the delay lines were considered. Special attention was given to explain how these factors the antenna properties. In the light of these studies, the following conclusions can be done:

i. **IMSL** type phase shifter is simple to dsign and provides the highest delay line efficiency η_{VDL} because ofthe low insertion loss of the microstrip line technology. Its tuning efficiency is low <0.8, and so, a long device in the range of $4\lambda_0$ to $6\lambda_0$ is needed to provide $\Delta\phi_{\beta,max} = -360°$. This also depends on the material tunability and the device geometry. Owing to the microstrip line technology, the structure can be meandered with a negligible performance reduction. Nevertheless, its response times are high, which is in a range of 1–3 m when the LC molecules are oriented by the alignment layer.

ii. **LC varactor loaded CPW** provides the highest η_{VDL} of the loaded line topologies as well as with the shortest physical length to achieve $\Delta\phi_{\beta,max} = 360°$. However, the device cannot be meandered by using the standard LC display technology, and therefore, the manufacturing costs are increased. Even if the device is bended, its performance reduces considerably.

iii. **LC varactor loaded slotline** topology enables to implement the varactors in parallel plate capacitor topology. Hence, they provide high tuning efficiency η_T and quality factor Q_{VAR} simultaneously. However, the challenge is to suppress the radiation losses with a proper substrate selection.

iv. **LC varactor loaded microstrip line** proposed in this work requires no vias through the glass substrate. Thus, the low-loss microstrip line technology can be used with a cost-effective way. The OSRR type varactors have the highest η_T because of its resonance behaviour. In contrary, Q_{VAR} is minimized, leading to a low measured η_{VDL}.

The delay lines proposed in this work are broadband devices, whereas they were customized for different operating frequencies. It should be mentioned, that any of those devices can be redesigned for another operating frequency as long as the LC technology is convenient. This means that there is no limiting factor defined by the device. The only critical factor is the high dielectric losses of the LC material at lower microwave frequencies.

As a result, the IMSL topology was chosen for the beam steering phased array. which was presented in the third section. The main reason for choosing the IMSL topology was its high delay line efficiency, even though it provided the highest response time. The response time of the delay line defines the beam steering rate of the antenna. The proposed antenna is well suited for portable terminals such as caravans, or for stationary terminals like satellite receptions at home.

The antenna was implemented into three dielectric layers with an overall thickness of 1.5 mm only. It included 2×2 microstrip patch antenna array, variable delay lines, dc-blocking structures, RF-feeding, and biasing networks. These components were also individually realized and measured. The IMSL was implemented first time with a spiral geometry, which enabled to design simple and compact RF-feeding network.

The measured antenna return losses were about 15 dB at the operating frequency of 17.5 GHz for different beam directions. Several measured far-field radiation patterns were presented for different beam directions and compared with the simulations. The cross polarization level and the antenna gain were measured as −16 dB and 6 dB, respectively, at the broadside.

References

[Af4] *AF45 Alkali Free Thin Glass* (Präzisions Glas and Optik GmbH, Iserlohn), http://www.pgo-online.com/intl/katalog/AF45.html

[BR08] N.S. Barker, G.M. Rebeiz, Distributed MEMS true-time delay phase shifters and wide-band switches. IEEE Trans. Microw. Theory Tech. **46**(11), 1881–1890 (1998)

[Bha06] A.K. Bhattacharyya, *Phased Array Antennas, Floquet Analysis, Synthesis, BFNs and Active Array Systems* (Wiley-Interscience, Wiley, New York, 2006)

[BP80] I.J. Bahl, P. Bhartia, *Microstrip Antennas* (Artech House, Norwood, 1980)

[BMS10] S. Bulja, D. Mirshekar-Syahkal, Meander line millimetre-wave liquid crystal based phase shifter. Electron. Lett. **46**(11), 769–771 (2010)

[Bal05] C.A. Balanis, *Antenna Theory: Analysis and Design*, 3rd edn. (Wiley, New York, 2005)

[Bor+00] A. Borgioli, Y. Liu, A.S. Nagra, R.A. York, Low-loss distributed MEMS phase shifter. IEEE Microw. Guided Wave Lett. **10**(1), 7–9 (2000)

[Cap09] F. Capolino (ed.), *Applications of Metamaterials*, Metamaterials Handbook (CRC Press, Taylor and Francis Group, Boca Raton, 2009)

[CVH08] K.V. Caekenberghe, T. Vaha-Heikkila, An analog RF MEMS slotline true-time-delay phase shifter. IEEE Trans. Microw. Theory Tech. **56**(9), 2151–2159 (2008)

[Coh69] S.B. Cohn, Slot line on a dielectric substrate. IEEE Trans. Microw. Theory Tech. **17**, 769–778 (1969)

[CS07] R.J. Collier, A.D. Skinner (eds.), *Microwave Measurements*, IET Electrical Measurement Series, 3rd edn. (The Institution of Engineering and Technology, London, 2007)

[Dol+93] D. Dolfi, M. Labeyrie, P. Joffre, J.P. Huignard, Liquid crystal microwave phase shifter. Electron. Lett. **29**, 926–928 (1993)

[DS+09] M. Duran-Sindreu, A. Velez, F. Aznar, G. Siso, J. Bonache, F. Martin, Applications of open split ring resonators and open complementary split ring resonators to the synthesis of artificial transmission lines and microwave passive components. IEEE Trans. Microw. Theory Tech. **57**, 3395–3403 (2009)

[Fal+04] F. Falcone, T. Lopetegi, J.D. Baena, R. Marin, F. Martin, M. Sorolla, Effective negative-ε stopband microstrip lines based on complementary split ring resonators. IEEE Trans. Microw. Wirel. Compon. Lett. **14**, 280–282 (2004)

[Fri+11] C. Fritzsch, S. Strunck, S. Bildik, R. Jakoby, Investigation of beam scanning speed for liquid crystal based tunable antennas, in *33rd ESA Antenna Workshop on Challenges for Space Antenna Systems* (2011)

[Fri94] D.A. Frickey, Conversions between S, Z, Y, H, ABCD, and T parameters which are valid for complex source and load impedances. IEEE Trans. Microw. Theory Tech. **42**, 205–211 (1994)

[Gar+12] Y. Garbovskiy, V. Zagorodnii, P. Krivosik, J. Lovejoy, R.E. Camley, Z. Celinski, A. Glushchenko, J. Dziaduszek, R. Dabrowski, Liquid crystal phase shifters at millimeter wave frequencies. J. Appl. Phys. **111**, 054504-1–054504-4 (2012)

[Gil+05] I. Gil, J. Bonache, J. Garcia-Garcia, F. Falcone, F. Martin, Metamaterials in microstrip technology for filter applications, in *Antennas and Propagation Society International Symposium, 2005 IEEE* (2005)

[Gei09] A. Giere, Material- und Bauteiloptimierung steuerbarer Mikrowellenkomponenten mit nichtlinearen Ferroelektrika. PhD thesis, Technische Universität Darmstadt, 2009

[Gil+11] M. Gil, C. Damm, M. Sazegar, M. Maasch, A. Giere, M. Schuessler, F. Aznar, A. Velez, J. Bonache, R. Jakoby, F. Martin, Tunable sub-wavelength resonators based on barium-strontium-titanate thick-film technology. IET Microw. Antennas Propag. **5**, 316–323 (2011)

[Gil+06] I. Gil, J. Bonache, J. Garcia-Garcia, F. Martin, Tunable metamaterial transmission lines based on varactor loaded split ring resonators. IEEE Trans. Microw. Theory Tech. **54**, 2665–2674 (2006)

[Gol01] M. Golio (ed.), *The RF and Microwave Handbook* , Electrical Engineering Handbook (CRC Press LLC., Boca Raton, 2001)

[Goe09] F. Gölden, Liquid Crystal Based Microwave Components with Fast Response Times: Material, Technology, Power Handling Capability. PhD thesis, Technische Universität Darmstadt, 2009

[Gup+96] K.C. Gupta, R. Garg, I. Bahl, P. Bhartia, *Microstrip Lines and Slotlines*, 2nd edn. (Artech House, Norwood, 1996)

[HY71] M. Hasegawa, M. Furukawa, H. Yanai, Slow wave propagation along a microstrip line on Si-SiO2 systems. Proc. IEEE **59**, 297–299 (1971)

[Het+00] K. Hettak, N. Dib, A. Sheta, A. Omar, G.-Y. Delisle, M. Stubbs, S. Toutain, New miniature broadband CPW-to-slotline transitions. IEEE Trans. Microw. Theory Tech. **48**, 138–146 (2000)

[HR03] J.S. Hayden, G.M. Rebeiz, Very low-loss distributed X-band and Ka-band MEMS phase shifters using metal-air-metal capacitors. IEEE Trans. Microw. Theory Tech. **51**(1), 309–314 (2003)

[HU+08] W. Hu, R. Cahill, R. Encinar, J.A. Dickie, H. Gamble, V. Fusco, N. Grant, Design and measurement of reconfigurable millimeter wave reflectarray cells with nematic liquid crystal. IEEE Trans. Microw. Theory Tech. **56**, 3112–3117 (2008)

[Jan+08] V. Janardhana, J.S. Roy, S. Pamidighantam, R.G. Kulkarni, Analysis of surface micro machined RF MEMS phase shifters. IEEE Aerosp. Electron. Syst. Mag. **23**(5), 32–35 (2008)

[KFN02] T. Kuki, H. Fujikake, T. Nomoto, Microwave variable delay line using dual-frequency switching-mode liquid crystal. IEEE Trans. Microw. Theory Tech. **50**, 2604–2609 (2002)

[KJDM78] R.W. Kreutel, D.F. DiFonzo Jr., C.E. Mahle, Satellite system measurements. Proc. IEEE **66**, 472–482 (1978)

[Kuk+02] T. Kuki, H. Fujikake, T. Nomoto, Y. Utsumi, Design of a microwave variable delay line using liquid crystal, and a study of its insertion loss. Electron. Commun. Jpn. (Part II: Electronics) **85**, 36–42 (2002)

[Kad+09] D. Kaddour, H. Issa, A.L. Franc, N. Corrao, E. Pistono, F. Podevin, J.M. Fournier, J.M. Duchamp, P. Ferrari, High-q slow-wave coplanar transmission lines on 0.35 μm CMOS process. IEEE Trans. Microw. Wirel. Compon. Lett. **19**(9), 542–544 (2009)

[Mal00] L.G. Maloratsky, Reviewing the basics of microstrip. Microwaves RF **39**(March), 79–88 (2000), http://www.lacoctelera.com/myfiles/ing-electronica/microstrip_basics.pdf

[MV09] P. Majumdar, A.K. Verma, Modified closed-form dispersion and loss models of slot-line with conductor thickness. J. Infrared Millim. Terahertz Waves **31**, 271–287 (2009)

[Mue07] S. Mueller. Grundlegende Untersuchungen steuerbarer passiver Flüssigkristall-Komponenten für die Mikrowellentechnik. PhD thesis, Technische Universität Darmstadt, 2007

[Mar+04] J. Martel, R. MarquTs, F. Falcone, J.D. Baena, F. Medina, F. Martfn, M. Sorolla, A new LC series element for compact bandpass filter design. IEEE Trans. Microw. Wirel. Compon. Lett. **14**, 210–212 (2004)

[Mar+03] F. Martin, J. Bonache, F. Falcone, M. Sorolla, R. Marques, Split ring resonator-based left-handed coplanar waveguide. Appl. Phys. Lett. **83**, 4652–4654 (2003)

[Muk+07] V.M. Mukhortov, S.I. Masychev, Y.I. Golovko, A.V. Chub, M. Mukhortov VI., A phase shifter on a slot line loaded with varactors designed on the basis on nanodimensional films of barium-strontium titanate. J. Commun. Technol. Electron. **52**, 1402–1406 (2007)

[Mir+08] I.G. Mironenko, A.A. Semenov, A.A. Ivanov, P.J. Beljavski, C.-Y. Kang, S.F. Karmanenko, Multislot transmission lines and microwave phase shifters based on (Ba, Sr)TiO3 films. Integr. Ferroelectrics Int. J. **97**(1), 58–66 (2008)

[NY99] A.S. Nagra, R.A. York, Distributed analog phase shifters with low insertion loss. IEEE Trans. Microw. Theory Tech. **47**(9), 1705–1711 (1999)

[Orb] Orbit/FR Inc. *Antenna Measurement Theory: Introduction to Antenna Measurement*, http://www.home.agilent.com/upload/cmc_upload/All/ORFR-Theory.pdf

[Pen+99] J.B. Pendry, A.J. Holden, D.J. Robbins, W.J. Stewart, Magnetism from conductors and enhanced nonlinear phenomena. IEEE Trans. Microw. Theory Tech. **47**, 2075–2084 (1999)

[PCS06] J. Perruisseau-Carrier, A.K. Skrivervik, Composite right/left-handed transmission line metamaterial phase shifters (MPS) in MMIC technology. IEEE Trans. Microw. Theory Tech. **54**(4), 1582–1589 (2006)

[Poz85] D.M. Pozar, A microstrip antenna aperture coupled to a microstrip line. Electron. Lett. **21**, 49–50 (1985)

[Poz11] D.M. Pozar, *Microwave Engineering*, 4th edn. (Wiley, New York, 2011)

[RSF05] R. Ramadoss, A. Sundaram, L.M. Feldner, RF MEMS phase shifters based on PCB MEMS technology. Electron. Lett. **41**(11), 654–656 (2005)

[Ra69] G.H. Robinson, J.L. Allen, Slot line application to miniature ferrite devices. IEEE Trans. Microw. Theory Tech. **17**, 1097–1101 (1969)

[RL01] I.D. Robertson, S. Lucyszyn (eds.), *RFIC and MMIC Design and Technology*, IET Circuits, Devices and Systems Series, vol. 13 (The Institution of Engineering and Technology, London, 2001)

[Rod+94] M.J.W. Rodwell, S.T. Allen, R.Y. Yu, M.G. Case, U. Bhattacharya, M. Reddy, E. Carman, M. Kamegawa, Y. Konishi, J. Pusl, R. Pullela, Active and nonlinear wave propagation devices in ultrafast electronics and optoelectronics. Proc. IEEE **82**, 1037–1059 (1994)

[Ras+00] J.-P. Raskin, G. Gauthier, L.P. Katehi, G.M. Rebeiz, Mode conversion at GCPW-to-microstrip-line transitions. IEEE Trans. Microw. Theory Tech. **48**, 158–161 (2000)

[SHK09] B. Sanadgol, S. Holzwarth, J. Kassner, 30 GHz liquid crystal phased array, in *Loughborough Antennas and Propagation Conference LAPC 2009* (2009)

[Saz+11a] M. Sazegar, A. Mehmood, Y. Zheng, H. Maune, Jak, Integrated resistive bias network for tunable devices on ferroelectric ceramics, in *Proc. of the 6th European Microwave Integrated Circuits Conference* (2011)

[Saz+11b] M. Sazegar, Y. Zheng, H. Maune, C. Damm, X. Zhou, J. Binder, R. Jakoby, Low-cost phased-array antenna using compact tunable phase shifters based on ferroelectric ceramics. IEEE Trans. Microw. Theory Tech. **59**, 1265–1273 (2011)

[Sio81] A. Sion, New method for direct G/T measurement using satellite signals. Electron. Lett. **17**, 883–885 (1981)

[Sim01] R.N. Simons, *Coplanar Waveguide Circuits, Components, and Systems* (Wiley, New York, 2001)

[Smi+00] D.R. Smith, W.J. Padilla, D.C. Vier, S.C. Nemat-Nasser, S. Schultz, Composite medium with simultaneously negative permeability and permittivity. Phys. Rev. Lett. **84**, 4184–4187 (2000)

[SCHc] SCHOTT, *Schott Borofloat 33*, http://www.schott.com/borofloat/english/download/borofloat_33_e.pdf

[VBM08] A. Velez, J. Bonache, F. Martin, Varactor-loaded complementary split ring resonators (VLCSRR) and their application to tunable metamaterial transmission lines. IEEE Trans. Microw. Wirel. Compon. Lett. **18**, 28–30 (2008)

[YW06] D.-K. Yang, S.-T. Wu, *Fundamentals of Liquid Crystal Devices* (Wiley, New York, 2006)

Chapter 5
Polarization Agile Antennas in LC Technology

A polarization agile antenna, for instance in a receiving mode, is able to adopt its polarization state to that of an incoming signal without changing its operating frequency. This increases the channel capacity since the frequency is reused by means of different polarizations. Similarly, depolarization losses can be compensated, which are caused by rain drops or occur in multipath propagation, especially in urban areas.

In this chapter, two polarization agile antennas are presented, using the liquid crystal (LC) technology for the first time. Section 5.1 focuses on the first antenna prototype, which includes a tunable coupled line. Its polarization state can be tuned between a linear polarization and a circular polarization by tuning the coupled line's coupling coefficient. This topology was proposed in [Fer+09] by using the semiconductor varactors for an S-band application. For higher operating frequencies, on the other hand, new technologies are required since the semiconductor based components are not adequate, because of their low performance. Hence, the proposed tunable coupled line is implemented by using the LC varactors in this work. A design algorithm is described and particularly, new methods are proposed to reduce the tunability required, which is necessary to control the coupling coefficient. Section 5.2 presents a second antenna array, which has an extended polarization capability than the first one. Its polarization state can be changed between dual-linear and dual-circular polarizations. This is accomplished by providing a differential phase shift $\Delta\phi_\beta$ between the two orthogonal feedings of the radiating elements. The radio frequency (RF)-feeding networks are implemented in inverted microstrip line topology (IMSL, see Sect. 2.2). Hence, no additional tunable component is necessary, indeed the RF-feeding networks provide a desired $\Delta\phi_\beta$. For the proof-of-concept, both prototypes are realized to operate at 13.75 GHz. Their measured reflection coefficients and far-field power patterns verify the proposed continuous polarization agility using the LC material.

O. H. Karabey, *Electronic Beam Steering and Polarization Agile Planar Antennas in Liquid Crystal Technology*, Springer Theses, DOI: 10.1007/978-3-319-01424-1_5,

5.1 Tunable Coupled Line Based Polarization Agile Antenna

5.1.1 Antenna Concept

The antenna prototype includes a dual-fed radiating element, which is connected to a tunable coupled line [Fer+09] as shown in Fig. 5.1. The radiating element could be either a dual-fed single element or a dual-fed array antenna. The tunability of the coupled line is achieved by loading the line with shunt LC varactors. The unloaded coupled line can be a forward or backward coupler whereas it operates always as a forward one when it is loaded. Hence, the RF signal fed from the input port (port 1) of the coupled line is delivered to the radiating structure through the direct (port 2) and coupled (port 4) ports. The isolation port (port 3) of the coupled line is terminated with a short end.

The polarization state of the antenna is controlled by tuning the varactors, which define the operating mode of the tunable coupled line. The device features two main operating modes:

i. *3-dB Hybrid Mode* (or *Coupled Mode*): In a transmitting antenna, for instance, all the energy fed from the input port is divided equally in magnitude and delivered to direct and coupled ports. Hence, the difference in the magnitude of the signals on these transmission ports is

$$\Delta|\underline{T}| =: |S_{21}| - |S_{41}| = 0\,\text{dB}. \tag{5.1}$$

At the same time, the phase difference of these signals is

$$\Delta\phi_T =: \angle S_{21} - \angle S_{41} = \pm 90°. \tag{5.2}$$

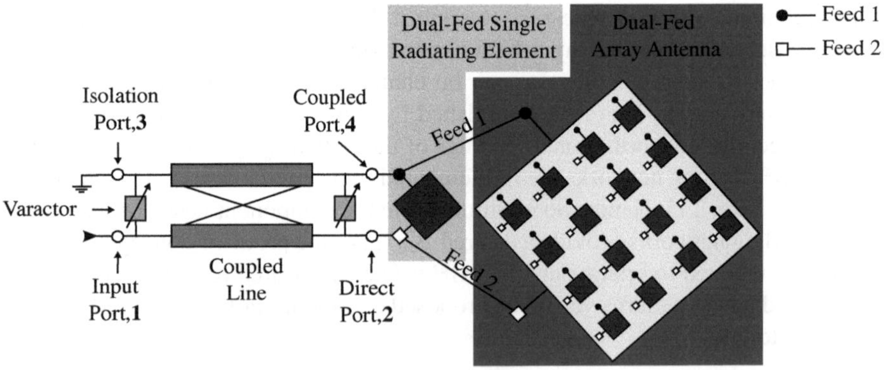

Fig. 5.1 Schematic of a loaded coupled line based polarization agile antenna. A tunable coupled line is connected to either a dual-fed single element or to a dual-fed array antenna

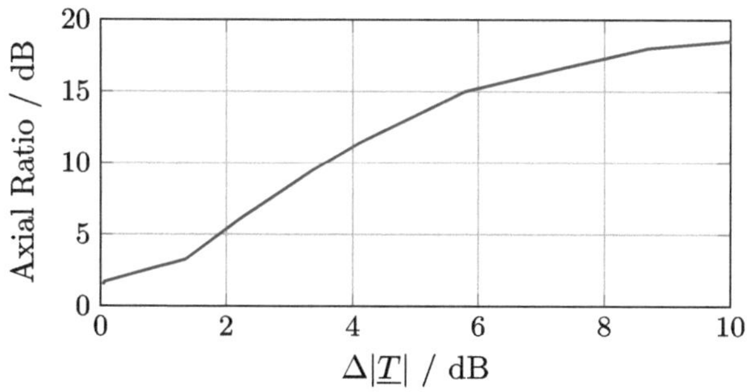

Fig. 5.2 Axial ratio of a dual-fed microstrip square patch antenna versus the magnitude difference of the transmission coefficients, $\Delta|\underline{T}|$

In this case, the antenna operates in a circular polarization, since its two orthogonal feedings receive identical signals with a $\pm 90°$ phase shift.

ii. *Uncoupled Mode*: All RF signal fed from the input port propagates only through the direct or only through the coupled port. This results in an antenna polarization of $+45°$ or $-45°$ linear polarization, respectively. In reality, however, there is always some amount of the power propagating through the undesired port. This results in an elliptical polarization with a high aspect ratio, which can still be treated as a linear polarization. For investigations, the axial ratio of a dual-fed microstrip square patch antenna is simulated versus $\Delta|\underline{T}|$ by using CST Microwave Studio. During the simulations, the circular polarization state of the antenna is enforced by keeping $\Delta\phi_T$ at $90°$ for different values of $\Delta|\underline{T}|$. However, when $\Delta|\underline{T}| = 10\,\text{dB}$, the axial ratio is about 18 dB at the broadside as shown in Fig. 5.2. This implies that the antenna polarization state is almost linear when there is $\pm 10\,\text{dB}$ difference in the magnitude of the transmitted signals.

Any elliptical polarization in between these two cases can be achieved as well, since the LC based varactors are continuously tunable. Additionally, if the signal is fed from the port 3 instead of port 1, i.e. by using an RF switch [Fer+09], the resultant polarizations will be orthogonal to the previous ones.

In the following section, a required effective varactor tunability τ_{eff} is determined, which is able to change $\Delta|\underline{T}|$ from 0 to $\pm 10\,\text{dB}$, or vice versa, leading to the antenna polarization agility circular and linear polarizations.

5.1.2 Tunable Coupled Line Analyzes

A 4-port *unit element* of the tunable coupled line, including a not tunable coupled line and shunt varactors, is shown in Fig. 5.3. The coupled line is represented by the characteristic impedance Z_{Line} and electrical length ϕ_ℓ. The varactor is shown only by its characteristic impedance Z_{VAR} that is calculated as $1/\omega C_0$ at a given angular

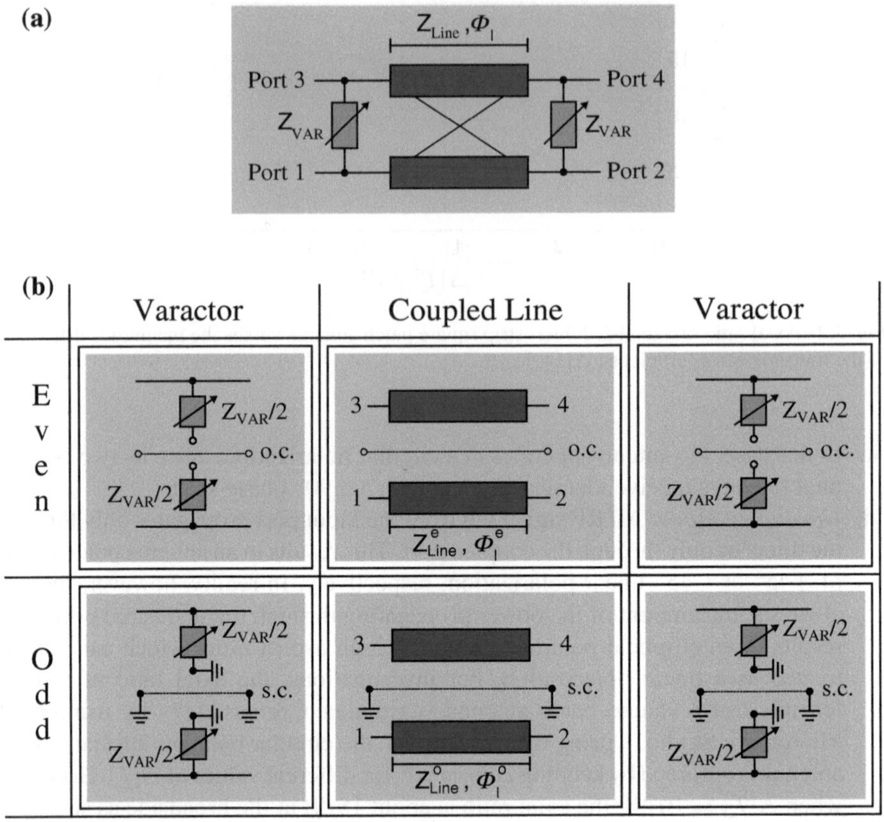

Fig. 5.3 **a** A unit element of the tunable coupled line. **b** Even and odd mode equivalent circuits of the cascaded components forming the unit element

frequency ω and capacitance C_0. For the analyzes, the transmission (ABCD) matrix approach is utilized since those matrixes of different components are cascaded as

$$A_{\mathrm{UE}} = A_{\mathrm{VAR}} \cdot A_{\mathrm{Coupled\ Line}} \cdot A_{\mathrm{VAR}}. \tag{5.3}$$

In Eq. (5.3), A_{VAR} and $A_{\mathrm{Coupled\ Line}}$ are the 4-port transmission matrixes of the varactor and the coupled line, respectively. Once A_{UE} is obtained, it can be converted into the scattering parameters by using the closed-form conversion formulas derived in Appendix A.5.

The network parameters are calculated by using the even and odd mode technique [Poz11]. This simplifies the 4-port network complexity to the 2-port one. In the even mode, no current follows through the transmission lines cutting the symmetry axis and therefore, they are terminated by open ends [Poz11]. In the odd mode, a voltage null occurs in the middle of these transmission lines. Hence, they are connected to a virtual ground in the equivalent circuit. These are practiced for the tunable coupled line, where the resultant equivalent circuits of its components are given in Fig. 5.3b.

The even (^{e}) and odd (^{o}) mode transmission matrixes of the varactor can be written straight forward as

$$A^e_{\text{VAR}} = \begin{pmatrix} 1 & 0 \\ 0 & 1 \end{pmatrix} \qquad A^o_{\text{VAR}} = \begin{pmatrix} 1 & 0 \\ \frac{2}{Z_{\text{VAR}}} & 1 \end{pmatrix}, \tag{5.4}$$

respectively. The even (odd) matrix of a lossless coupled line is given by [Poz11]

$$A^{e/o}_{\text{Coupled Line}} = \begin{pmatrix} \cos\phi^{e/o}_\ell & jZ^{e/o}_{\text{Line}} \sin\phi^{e/o}_\ell \\ \frac{j\sin\phi^{e/o}_\ell}{Z^{e/o}_{\text{Line}}} & \cos\phi^{e/o}_\ell \end{pmatrix}, \tag{5.5}$$

where Z^e_{Line} (Z^o_{Line}) is the characteristic impedance and ϕ^e_ℓ (ϕ^o_ℓ) is the electrical length of the even (odd) mode. These parameters can be determined accurately by using the formulas, e.g. presented in [ARJ75] for a given device geometry. However, since the device geometry is not defined for the moment, the even (odd) mode electrical parameters are calculated by using the approximations given in [Poz11] as

$$\phi^e_\ell = \phi^o_\ell = \phi_\ell, \tag{5.6}$$

$$Z^e_{\text{Line}} = Z_{\text{Line}}\sqrt{\frac{1+c}{1-c}}, \tag{5.7}$$

$$Z^o_{\text{Line}} = Z_{\text{Line}}\sqrt{\frac{1-c}{1+c}}, \tag{5.8}$$

where c is the midband ($\phi_\ell = \pi/2$) coupling coefficient of the not tunable coupled line. Here, $\phi^e_\ell = \phi^o_\ell$ refers that the both modes have identical phase velocities. Indeed, this deviates in practice depending on the device geometry and leads to a poor coupled line directivity [Poz11]. Nevertheless, this is a reasonable simplification for a narrowband design and can be further applied for broadband designs by using a phase velocity compensation technique [SS74, MMH92]. Additionally, if $\phi_\ell \neq \pi/2$ the coupling coefficient changes slightly, which can be recalculated by using the formulas given in [KJ84].

Once, the 2-port even and odd mode transmission matrixes are determined as in Eqs. (5.4) and (5.5), the resultant 4-port matrixes are calculated as [ZJ69]

$$\begin{bmatrix} V_1 \\ V_3 \\ I_1 \\ I_3 \end{bmatrix} = \begin{bmatrix} a_{11} & a_{12} & a_{13} & a_{14} \\ a_{21} & a_{22} & a_{23} & a_{24} \\ a_{31} & a_{32} & a_{33} & a_{34} \\ a_{41} & a_{42} & a_{43} & a_{44} \end{bmatrix} \cdot \begin{bmatrix} V_2 \\ V_4 \\ -I_2 \\ -I_4 \end{bmatrix} \tag{5.9}$$

with

$$\begin{aligned}
a_{11} = a_{22} = a_{33} = a_{44} &= \frac{1}{2}\left(A^e_{11} + A^o_{11}\right),\\
a_{12} = a_{21} = a_{34} = a_{43} &= \frac{1}{2}\left(A^e_{11} - A^o_{11}\right),\\
a_{13} = a_{24} &= \frac{1}{2}\left(A^e_{12} + A^o_{12}\right),\\
a_{14} = a_{23} &= \frac{1}{2}\left(A^e_{12} - A^o_{12}\right),\\
a_{31} = a_{42} &= \frac{1}{2}\left(A^e_{21} + A^o_{21}\right),\\
a_{32} = a_{41} &= \frac{1}{2}\left(A^e_{21} - A^o_{21}\right).
\end{aligned} \tag{5.10}$$

Hence, the transmission matrix of the varactor A_{VAR} is determined as

$$A_{\mathrm{VAR}} = \frac{1}{Z_{\mathrm{VAR}}} \cdot \begin{bmatrix} Z_{\mathrm{VAR}} & 0 & 0 & 0 \\ 0 & Z_{\mathrm{VAR}} & 0 & 0 \\ 1 & -1 & Z_{\mathrm{VAR}} & 0 \\ -1 & 1 & 0 & Z_{\mathrm{VAR}} \end{bmatrix} \tag{5.11}$$

by substituting the matrixes in Eq. (5.4) into Eq. (5.10). Similarly, $A_{\mathrm{Coupled\ Line}}$ is equal to

$$A_{\mathrm{Coupled\ Line}} = \frac{\sin\phi_\ell}{2} \begin{bmatrix} 2\cot\phi_\ell & 0 & j(Z^e_{\mathrm{Line}} + Z^o_{\mathrm{Line}}) & j(Z^e_{\mathrm{Line}} - Z^o_{\mathrm{Line}}) \\ 0 & 2\cot\phi_\ell & j(Z^e_{\mathrm{Line}} - Z^o_{\mathrm{Line}}) & j(Z^e_{\mathrm{Line}} + Z^o_{\mathrm{Line}}) \\ j(\frac{1}{Z^e_{\mathrm{Line}}} + \frac{1}{Z^o_{\mathrm{Line}}}) & j(\frac{1}{Z^e_{\mathrm{Line}}} - \frac{1}{Z^o_{\mathrm{Line}}}) & 2\cot\phi_\ell & 0 \\ j(\frac{1}{Z^e_{\mathrm{Line}}} - \frac{1}{Z^o_{\mathrm{Line}}}) & j(\frac{1}{Z^e_{\mathrm{Line}}} + \frac{1}{Z^o_{\mathrm{Line}}}) & 0 & 2\cot\phi_\ell \end{bmatrix}. \tag{5.12}$$

For empirical studies, a MatLab script has been written based on these formulas. The script determines a $Z_{\mathrm{VAR}} = 1/\omega C_0$ for a given coupling coefficient c and an electrical length ϕ_ℓ. The calculated Z_{VAR} values lead to the 3-dB hybrid operating mode where $\Delta|\underline{T}| = 0\,\mathrm{dB}$ and $\Delta\phi_T = \pm 90°$. The results and the magnitudes of the resultant reflection and isolation coefficients are depicted in Fig. 5.4 for one unit element ($N = 1$). During the calculations, Z_{Line} is specified to be $50\,\Omega$, which is a practical requirement.

In a next step, the pre-calculated varactor impedances are tuned to change the operating mode of the coupled line into the uncoupled mode, where $\Delta|\underline{T}| \geq \pm 10\,\mathrm{dB}$. As in the previous chapter, the varactor tuning refers to an effective tunability that is

$$\tau_{\mathrm{eff}} = \frac{C_{\max} - C_{\min}}{C_{\max}}, \tag{5.13}$$

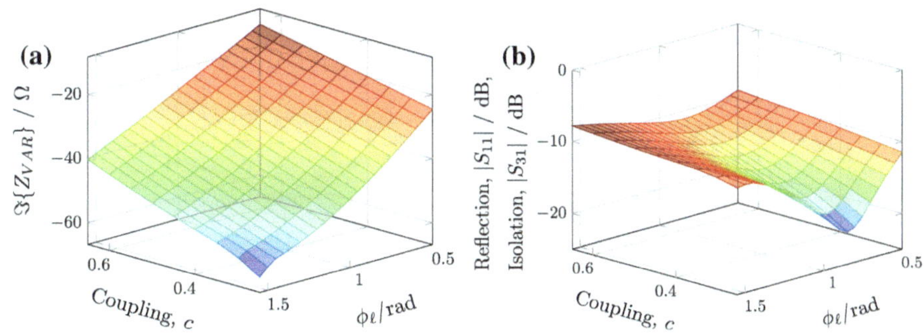

Fig. 5.4 **a** Required varactor impedances for a 3-dB hybrid mode operation versus the coupling coefficient c and coupled line electrical length ϕ_ℓ. **b** Magnitude of the reflection and isolation coefficients when the tunable coupled line operates in the 3-dB hybrid mode

where C_{max} and C_{min} are the maximum and minimum capacitances, respectively, achieved by the varactor tuning. It should be noted that a varactor can be tuned in two ways. First, the capacitance resulting the 3-dB hybrid mode C_0 can be assumed as C_{min} and then C_{max} is calculated for the given τ_{eff}. Second, C_0 is assumed to be equal to C_{max} and then C_{min} is calculated. The first one is so called as *forward tuning* since the capacitance is increased by the tuning. Similarly, the second one is so called as *back tuning*.

Figure 5.5, shows $\Delta|\underline{T}|$ and $\Delta\phi_T$ when the forward and back tunings are performed when $N = 1$ and $N = 5$. This empirical study leads to following interpretations:

i. The coupling coefficient c of the not tunable coupled line hardly affects the performance of the tunable coupled line. Actually, c has an impact on the selection of a Z_{VAR} value to operate in the 3-dB hybrid mode according to Fig. 5.4. Once this operating mode is satisfied, the performance of the tunable coupled line is sensitive only to an applied tunability. Hence, a convenient Z_{VAR} and c pair can be chosen depending on the device geometry without effecting the device performance.
ii. The tunability required is reduced when the unit element number N is increased. In the back tuning case, for instance, a tunability of 0.6 is required to changed $\Delta|\underline{T}|$ from range 0 to 10 dB when $N = 1$. On the other hand, the tunability required reduces to 0.1 when $N = 5$. The same holds true for the forward tuning.
iii. The tunability required is again reduced when the forward tuning is performed. This can be explained intuitively as: the tunable coupled line is a low pass device since it is loaded with shunt capacitive loads. As the forward tuning is performed, its cut-off frequency reduces because the capacitance of the shunt loads increases. Hence, the cut-off frequency of the device approaches to a given operating frequency and so, its tuning efficiency increases like the varactors presented in the previous chapter.

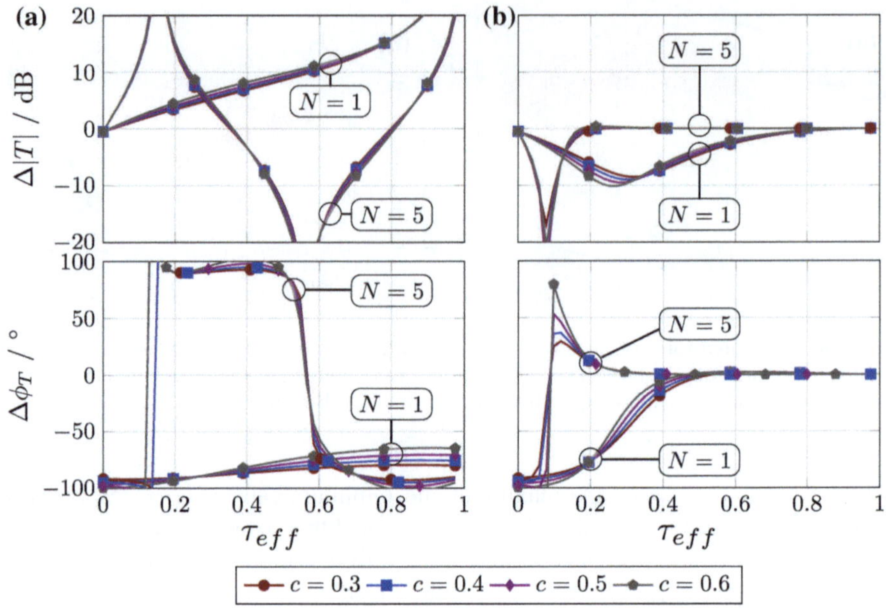

Fig. 5.5 Difference in the magnitude $\Delta|\underline{T}|$ and phase $\Delta\phi_T$ of the signals on the transmission ports. **a** The back and **b** forward tunings are performed for different coupling coefficients c when $\phi_\ell = \pi/2$. N is the unit element number

iv. There are several solutions, where the polarization state of the antenna can be changed, if the element number is increased and the back tuning is performed. Table 5.1 lists these solutions according to Fig. 5.5. These results show that the polarization state of the antenna configuration given in Fig. 5.1 can be controlled between dual linear and dual circular polarizations if the unit elements are cascaded and the back tuning is performed

5.1.3 Realization and Measurement Results

A polarization agile antenna prototype is realized and measured to verify the investigations described in the previous section. The prototype, which is shown in Fig. 5.6, is designed to operate at 14 GHz. It consists of a tunable coupled line with five unit elements ($N = 5$) and a dual-fed microstrip patch antenna. The LC material is contained between two glass substrates, which are so called as the back and front substrates (Fig. 5.6b), respectively. Underlying substrate details and the layout dimensions are given in Table 5.2 and in Appendix A.4.5, respectively.

The coupled line is implemented in microstrip line technology. As shown in Fig. 5.6b, the ground electrode is patterned on the bottom side of the back substrate by

Table 5.1 Achievable antenna polarization states when the coupled line unit element number $N = 5$ and the back tuning is performed

τ_{eff}	$\Delta\lvert\underline{T}\rvert$ (dB)	$\Delta\phi_T$	Antenna polarization state
0	0	−92°	Case I : Circular polarization
0.1	10	−100°	Case II : Linear polarization
0.35	0	95°	Case III : Circular polarization (Reverse of Case I)
0.48	−10	100°	Case IV : Linear polarization (Orthogonal to Case II)

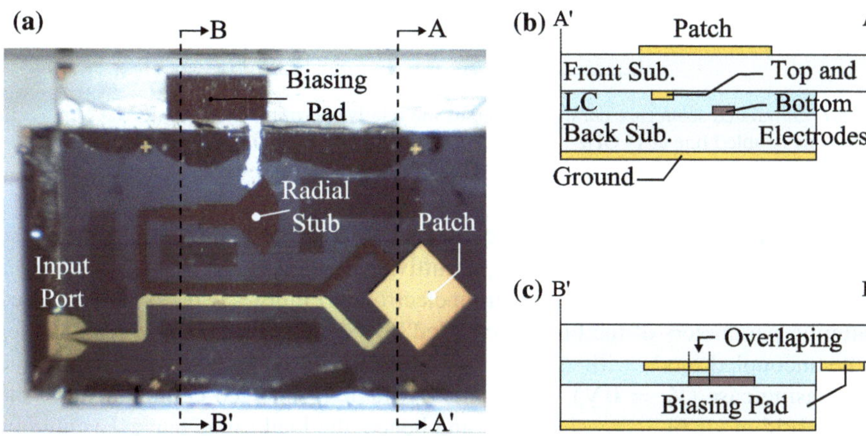

Fig. 5.6 **a** A photograph and **b**–**c** two cross sectional views of the antenna prototype. The isolation port of the tunable coupled line is shorted by a radial stub

Table 5.2 Details of the substrates used in the polarization agile antenna element

	Type	ε_r	tan δ	Thickness (μm)
Glass substrate	Fused silica	3.78	0.0015	300
LC layer	TUD-566	2.32–3.11	0.0066–0.0021	10

using a standard photolithography technique (see Sect. 2.2.1). The signal electrodes of the coupled line are fabricated symmetrically on the top side of the back substrate and on the bottom side of the front substrate. These electrodes include finger-like structures extending from one side. The electrodes are covered by the alignment layer to orient the LC molecules parallel to the surfaces. The tunable coupled line is actually formed when the substrates are aligned and stacked in such a way that the electrodes are on the inner sides and have a contact with the LC layer. Then, the fingers overlap and the overlapping areas form the varactors in the parallel plate capacitor topology as shown in Fig. 5.6c. The dual-fed microstrip patch antenna is located on the top side of the front dielectric. The RF signal is coupled between the tunable coupled line and the microstrip patch by using the proximity coupling technique [Bal05].

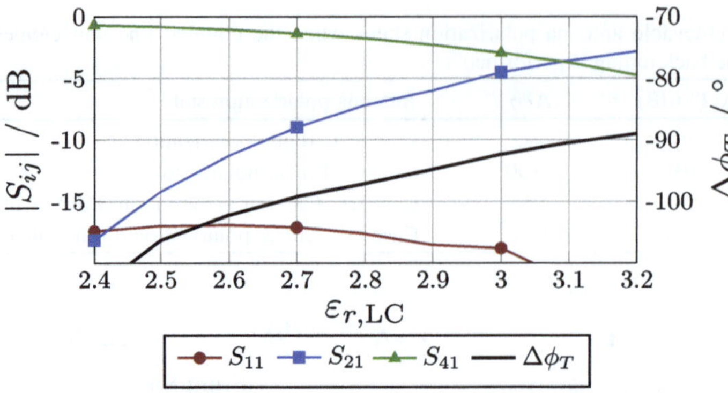

Fig. 5.7 Simulated scattering parameters versus the relative permittivity of the LC layer $\varepsilon_{r,LC}$ of the tunable coupled line at 14 GHz

To achieve a desired operating mode of the coupled line, the orientation of the LC layer, and therefore, its relative permittivity $\varepsilon_{r,\text{LC}}$ is controlled by applying a bias voltage V_b across the top and bottom electrodes. Figure 5.7 shows the simulated scattering parameters of the tunable coupled line versus $\varepsilon_{r,\text{LC}}$. The device operates in the uncoupled mode with $\Delta|\underline{T}| = -17$ dB when $\varepsilon_{r,\text{LC}} = 2.4$, which refers the zero biasing case ($V_b = 0$ V). The operating mode is changed continuously to the 3-dB hybrid mode by increasing $\varepsilon_{r,\text{LC}}$ and hence, the load capacitance. Eventually, the 3-dB hybrid mode is achieved when $\varepsilon_{r,\text{LC}} = 3.1$. These results show that the back tuning is performed since the operating mode can be changed from the 3-dB hybrid mode to the uncoupled mode by reducing the capacitance. Additionally, it is enough to tune $\varepsilon_{r,\text{LC}}$ from 2.6 to 3.1 to change $\Delta|\underline{T}|$ from -10 to 0 dB. This leads to a material tunability τ_{LC} of 0.161 (see Eq. 2.19). According to Table 5.1, on the other hand, the effective tunability τ_{eff} is supposed to be 0.1 for the same functionality. As described in Sect. 4.2.2, the ratio between τ_{eff} and τ_{LC} gives the tuning efficiency of the varactor η_τ, which is equal to 0.62 in this case.

The reflection coefficients of the antenna prototype are measured for different bias voltages using an on-wafer probe station. These results are given in Fig. 5.8. The results are emphasized for the zero biasing and the full biasing ($V_b = 40$ V) cases. The antenna operates at 13.75 GHz with a return loss of better than 10 dB. Its 10 dB fractional bandwidth is reduced from 8.1 to 4 %. when V_b is applied from 0 to 40 V. 4 % bandwidth is low but typical for the microstrip patch antennas. However, a reduction of the bandwidth has to be prevented because neither antenna's operating frequency nor its bandwidth should be changed while its polarization is tuned. This could be solved by using an aperture coupling instead of the proximity one.

Figure 5.9 shows the normalized received powers of two different setups, which are measured in an anechoic chamber. Linearly polarized signals are received by the prototype for different biasing states while radiating in the broadside direction. In the first setup, the prototype is located in such a way that a co-polarization (CoPol) is achieved with respect to the feed 1 (see Fig. 5.1). In the zero biasing case, the

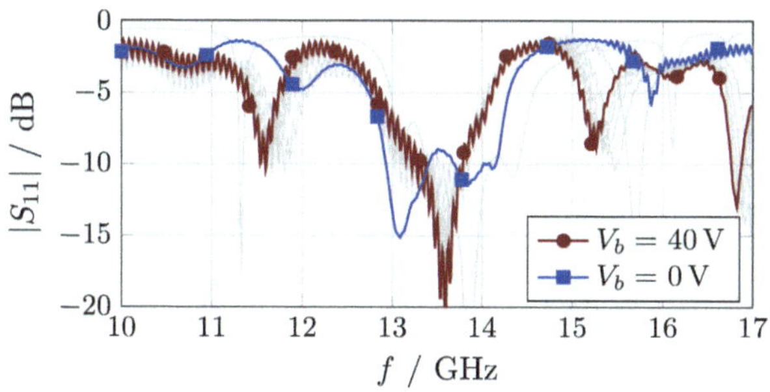

Fig. 5.8 Measured reflection coefficients of the antenna prototype

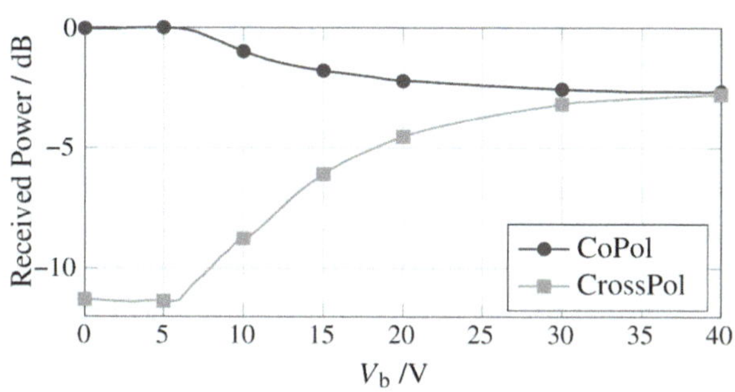

Fig. 5.9 Measured received powers of the antenna at the broadside versus the applied bias voltage V_b. The curves are normalized with respect to the maximum received power. The co-and cross-polarizations are defined with respect to the feed 1

antenna is linearly polarized through the feed 1, and therefore, the received power is maximized, since the polarizations of the antenna and the received signal are matched. The received power decreases continuously as the applied voltage increases. The minimum received power is measured at the full biasing case and it is 3 dB less than the highest value. The reason is that the prototype operates in a circular polarization at the full biasing case and the received signal is linearly polarized. These results are confirmed in a second setup, in which the only difference is that the prototype is turned 90° mechanically. Hence, a cross-polarization (CrossPol) can be measured with respect to feed 1. In contrast to the first measurements, the received power is minimized at the zero biasing state because the polarization of the received signal is orthogonal to the polarization of the prototype at this state. However, the received power increases continuously with the applied bias voltage. At the full biasing state, for instance, the maximum power is measured and it is equal to the one measured from the first setup since the prototype operates in the circular polarization.

Finally, the measured far-field power patterns of the zero and full biasing states are given for other azimuth angles in Fig. 5.10. A CrossPol level of −12 dB is measured when $V_b = 0$ V. At the full biasing case, on the other hand, both curves coincide at the broadside direction since the antenna operates in the circular polarization state.

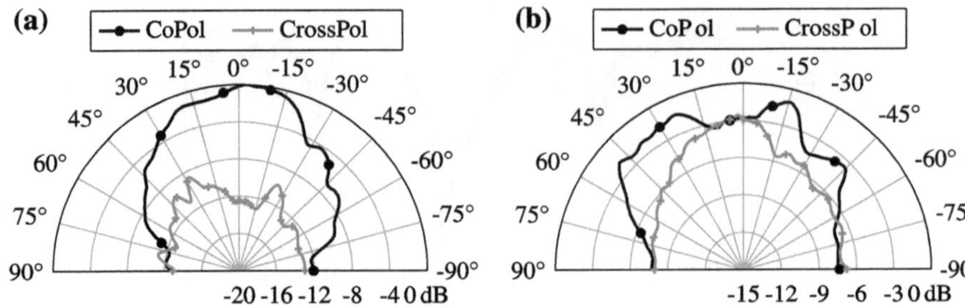

Fig. 5.10 Measured far-field received powers of the prototype. **a** The zero biasing case ($V_b = 0\,\text{V}$), where the antenna polarization is linear. **b** The full biasing case ($V_b = 40\,\text{V}$) and the antenna operates in the circular polarization state. The co- and cross-polarizations are defined with respect to the feed 1. The measured raw data are normalized and smoothed by using MatLab 5th order moving average function

It should be noted that, their half power beam widths are different at large azimuth angles. This is a typical behaviour of circularly polarized square patch antenna, which indicates that the axial ratio is minimizes at the broadside and arises at large azimuth angles [Bal05]. Nevertheless, the axial ratios are not stated quantitatively herein, since the measurements are performed by using a linearly polarized transmitting antenna.

5.2 Variable Delay Line Based Polarization Agile Antenna

5.2.1 Antenna Concept

This section presents another polarization agile planar antenna, which is designed to operate at 13.75 GHz by using the LC based variable delay lines. The antenna includes a 2×2 dual-fed microstrip patch antenna array and two separate RF-feeding networks combining each vertical and horizontal feedings of the dual-fed antenna separately. The operation principle of the antenna is given in Fig. 5.11 [GSZ06]. Its polarization state can be controlled continuously between dual linear and dual circular polarizations. This is accomplished by introducing a differential phase shift $\Delta\phi_\beta$ between the two orthogonal antenna feedings. According to Fig. 5.11b, the required $\Delta\phi_\beta$ is in the range of 0–180°. The proposed antenna needs no additional tunable device to provide the required $\Delta\phi_\beta$. Indeed, the RF-feeding networks are implemented in the inverted microstrip line (IMSL) topology and therefore, a differential phase shift is obtained by tuning the LC material. Additionally, any polarization state between the circular and linear ones are achievable like in the previous antenna because of the continuous tunability of the LC material.

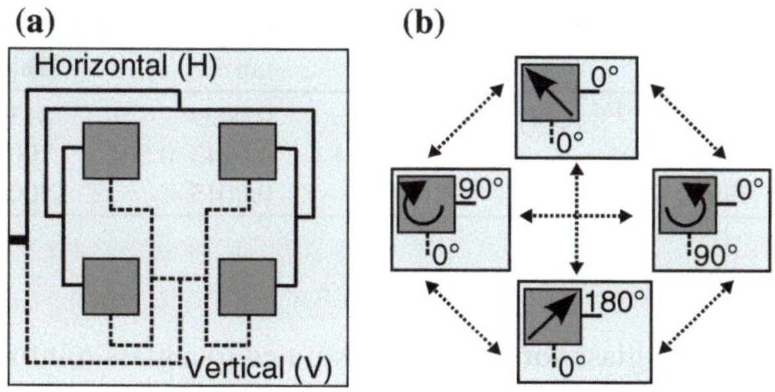

Fig. 5.11 Operation principle of a polarization agile antenna integrated with a tunable phase shifter/delay line. **a** The layout includes two RF-feeding networks for the vertical and horizontal feedings. **b** The polarization states are illustrated for +45° linear (*top*), left-/right-handed circular polarizations (*middle*) and −45° linear polarizations (*bottom*)

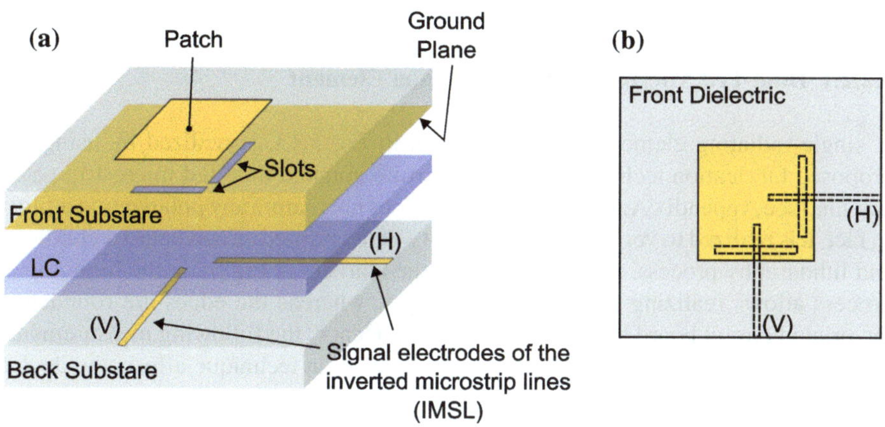

Fig. 5.12 Schematics of the single radiating element. **a** Exploded perspective and **b** Bottom views

5.2.2 Design and Verification

Schematics of the proposed antenna are given for a single radiating element in Fig. 5.12. The LC material is encapsulated between the front and back substrates. The dual-fed microstrip patch antenna and its ground plane are implemented on the top and bottom sides of the front dielectric, respectively. The ground plane includes the slots underlying the patches to form an aperture coupling between the antenna and RF-feeding networks. The electrodes of the two feeding networks are located on the top side of the back substrate.

For an LC device realization, glass substrates are mainly preferred because of their mechanically stability and transparency. However, it is an issue to pattern the both sides of a glass substrate by using the in-house facilities according to experiences gained in the phased array realization (see Sect. 4.3). Hence, for a cost-effective and

Table 5.3 Details of the substrates used in the polarization agile antenna element

	Type	ε_r	$\tan\delta$	Thickness (μm)
Front substrate	RO4003C^{TM} [Cor]	3.38	0.0027	813
LC layer	GT3-23001	2.46–3.28	0.0143–0.004	100
Back substrate	Fused silica [SCHd]	3.78	0.0015	300

simpler fabrication, the glass substrate is substituted to a glass reinforced hydrocarbon/ceramic material from Rogers Corporation for the front substrate. The electrodes are patterned by using a rapid printed circuit board prototyping machine. For the back substrate, on the other hand, still a glass substrate is used since it allows aligning and stacking the front and back substrates. The cavity between the two substrates is achieved by using micro pearls. The details of the substrates are given in Table 5.3.

5.2.2.1 Dual-Fed Microstrip Patch Antenna Element

A single radiating element, which is shown in Fig. 5.13, is realized by using the proposed fabrication technique. It is an aperture coupled dual-fed microstrip patch antenna (see Appendix A.4.6). This structure does not feature any polarization agility. In fact, it is realized to verify the compatibility of the printed circuit board technology and lithography process. As can be seen in the markers in Fig. 5.13, the lithography process allows realizing sharp electrode edges, whereas the edges are rounded if the printed circuit board technology is utilized. Hence, the following measurements are performed to verify whether or not this fabrication technique affects the device performance.

The reflection coefficients of the structure are measured by using the on-wafer probe station with a ground-signal-ground probe. During the measurements, a dc

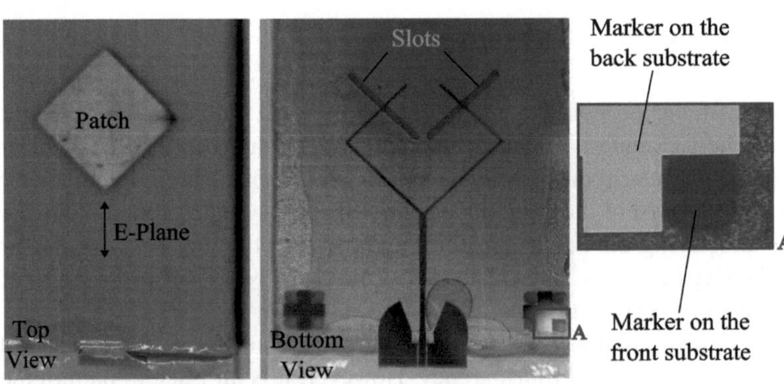

Fig. 5.13 Photographs of the realized dual-fed microstrip patch antenna element by using the proposed cost-effective fabrication process

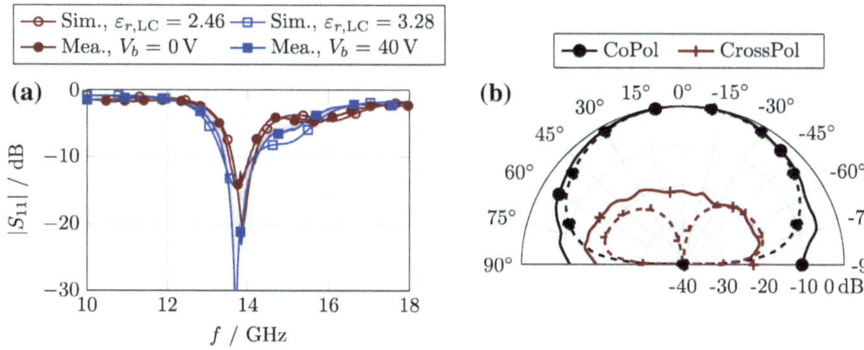

Fig. 5.14 **a** Measured and simulated reflection coefficients of the single radiating element. The zero biasing ($V_b = 0\,\mathrm{V}$) and full biasing ($V_b = 40\,\mathrm{V}$) states refer to the lowest and highest values of the relative permittivity of the LC material $\varepsilon_{r,LC}$, respectively. **b** Normalized far-field power pattern measurement results for co- and cross-polarizations. *Solid* and *dashed lines* represent the measured and simulated results, respectively

biasing voltage V_b is applied across the ground and IMSL signal electrodes. The bias voltage is changed from 0 to 40 V in step of 5 V, with a leakage current of less than 50 nA. The simulated and measured reflection coefficients are given in Fig. 5.14a. The antenna operates at 13.75 GHz with a return loss of better than 15 dB. Its operating frequency is hardly affected by applying a bias voltage because the aperture coupling technique is used. As a result, the 10 dB fractional bandwidth varies slightly from 5.5 to 6 % depending on the biasing voltage. Additionally, far-field power patterns of the single element are measured as well in the anechoic chamber at the operating frequency. These results are given in Fig. 5.14b for the co- and cross-polarizations. The antenna has a cross polarization level of 15–20 dB.

These results lead to the conclusion that the proposed fabrication technique can be used for cost-effective realizations of planar LC antennas, since both, reflection and far-field pattern measurement results match quite well to the simulation results.

5.2.2.2 RF-Feeding Networks

Layout of the RF-feeding networks are given in Fig. 5.15 for a 2 × 2 antenna array. These electrodes are located on the top side of the back substrate. The lines are meandered to implement the RF-feeding network into the limited area that is $0.46\lambda_0 \times 0.46\lambda_0$ at 13.75 GHz. The layout also includes interdigital capacitors (IDCs) in the main branches as being a contactless RF interconnect. Hence, any dc contact between the two RF-feeding networks is avoided to bias them separately. The impedance matching is carried out using the standard quarter-wave impedance transformers.

The RF-feeding networks deliver the signal to the radiating elements equally in magnitude and in phase through the V and H feedings. Simulated scattering parameters of the RF-feeding networks are plotted in Fig. 5.16. According to the given

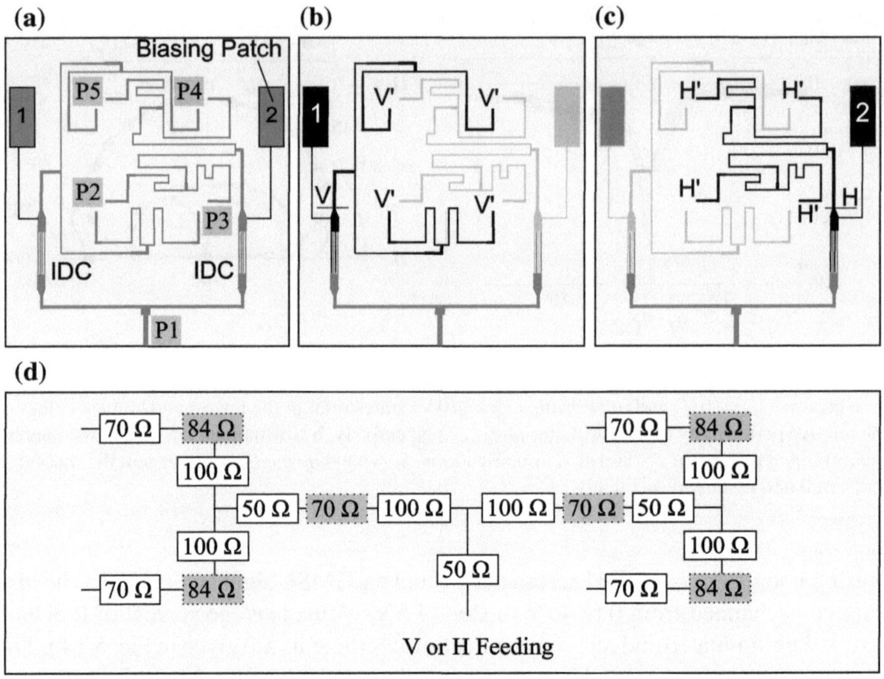

Fig. 5.15 **a** A complete layout of the RF-feeding networks, including the interdigital capacitors (IDCs). P1 to P5 shows the port definitions used in the followings. **b** RF-Feeding networks highlighted for the vertical (V) and **c** the horizontal (H) excitations of the radiating elements. **d** A schematic showing the characteristic impedances of the transmission lines. The 70 and 84 Ω sections are the quarter-wave impedance transformers

layout, the signal is divided three times while propagating from the input port to any of the output ports. Thus, the magnitude of the transmission coefficients are expected to be −9 dB for lossless lines. However, the transmission coefficients plotted in Fig. 5.16 are around −11.5 dB in average. This implies that the insertion losses caused by the RF-feeding networks are 2.5 dB. The curves fluctuate since the lines are meandered several times. The fluctuations are less than is 1.5 dB for the worst case. This only effects the axial ratio of the antenna polarization and does not change the operating polarization state. This is shown implicitly in Fig. 5.2. A constant signal magnitude can be obtained by meandering the lines with a proper line geometry. Additionally, results in Fig. 5.16 verify that the phases of the output signals match quite well to each other at different tuning states. This is essential to achieve since the antenna polarization is defined by the differential phase shift between the RF-feeding networks.

The physical length of the IMSL line is supposed to be 50 mm to obtained the $\Delta\phi_{\beta,\max} = 180°$ at 13.75 GHz [Mue07]. The proposed RF-feeding networks satisfy this requirement (see Fig. 5.16). However, the physical length of the feeding lines

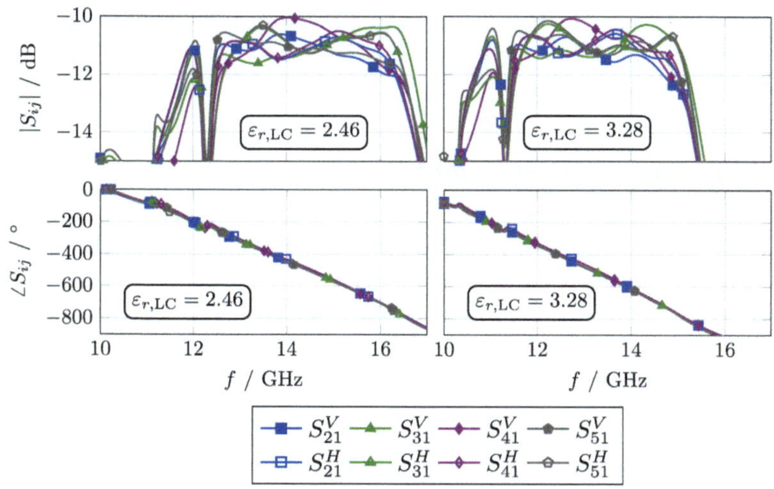

Fig. 5.16 Simulated scattering parameters of the RF-feeding networks for different relative permittivities of LC layer $\varepsilon_{r,LC}$. The port definitions are given in Fig. 5.15a. *V* and *H* stands for the vertical and horizontal feedings, respectively

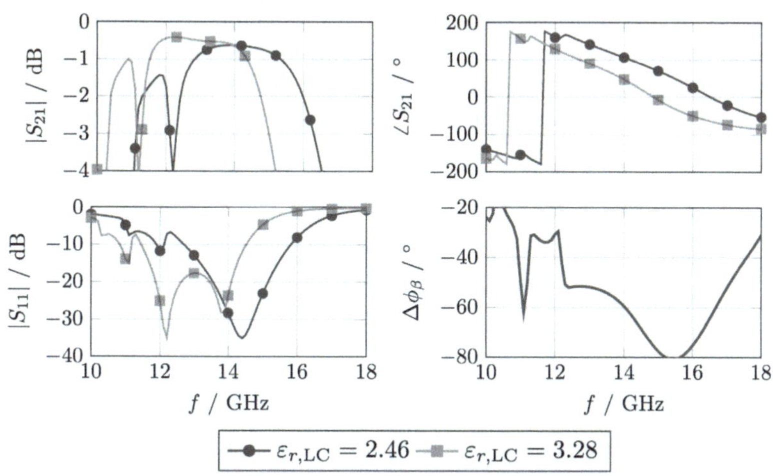

Fig. 5.17 Simulated scattering parameters of the interdigital capacitors (IDCs) for different relative permittivities of LC layer $\varepsilon_{r,LC}$

between V-V' and H-H' in Fig. 5.15b, c, respectively, are reduced to 37 mm, which results in $\Delta\phi_{\beta,max}$ of 130°. The remaining $\Delta\phi_{\beta,\max}$ is provided by the IDCs. The simulation results given in Fig. 5.17 show that the IDCs feature a $\Delta\phi_{\beta,\max} = 56°$ at the operating frequency with an 0.65 dB insertion loss.

The bias voltage V_b is applied through the biasing patches and delivered to the RF circuitry via bias lines. These lines are implemented in the adhesive chromium layer between the gold electrodes and the glass substrate [Saz+11b]. The resultant dc resistance between the biasing patch and the signal electrode of the IMSL is calculated to be 4.7 kΩ.

5.2.3 Measurement Results and Discussions

The realized polarization agile antenna is shown in Fig. 5.18. Its simulated and measured reflection coefficients are plotted in Fig. 5.19.

In the zero biasing case, the antenna features a return loss of better than 10 dB in a frequency range of 13.3–16 GHz. Increasing the bias voltage decreases the center frequency slightly. Especially, when $V_b = 40$ V, the -10 dB bandwidth is measured from 13.5 to 15 GHz.

The far-field power pattern of the antenna is measured in the anechoic chamber. The two measurement processes explained in Sect. 5.1.3 are performed herein. These results are plotted in Fig. 5.20. In the first setup, a co-polarization is achieved along the +45° axis (see Fig. 5.18) and the bias voltage is varied. In the second setup, the same procedure is repeated after a co-polarization along the −45° is obtained. At the zero biasing and the full biasing states, the antenna operates in linear polarizations along +45° and −45° axes, respectively. The corresponding cross-polarization levels are approximately 9 dB for the former and 6 dB for the latter. Additionally, the both curves coincide at $V_b = 7.5$ V, which confirms that the antenna operates at a circular polarization state at this bias voltage.

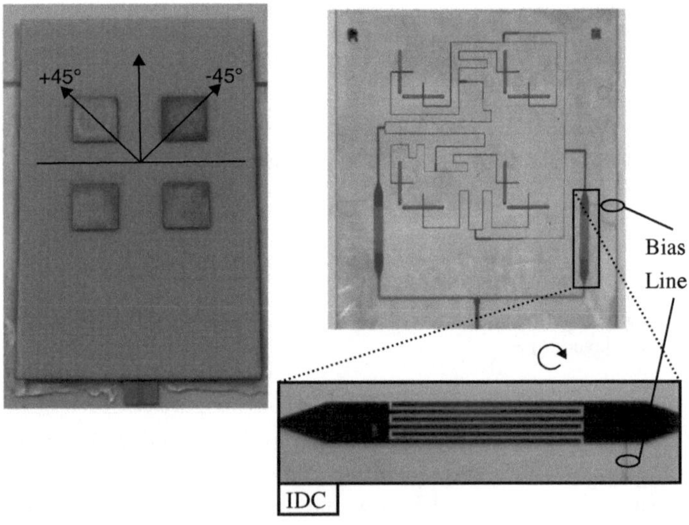

Fig. 5.18 *Left* Photographs of the polarization agile antenna from the *top view*. *Right* The picture is taken from the *bottom views*

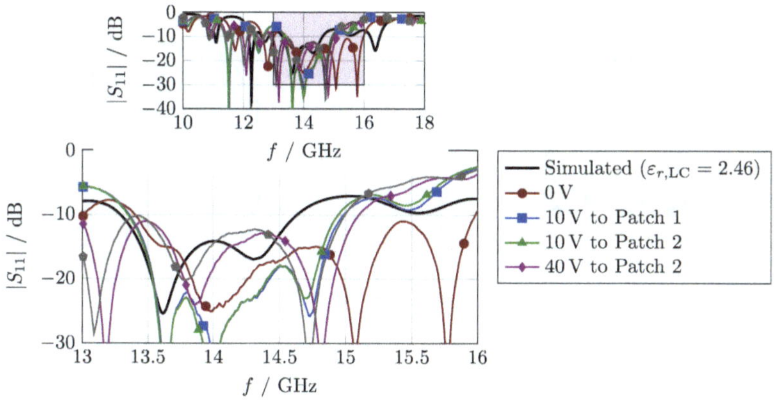

Fig. 5.19 Measured reflection coefficients of the antenna array at different biasing states

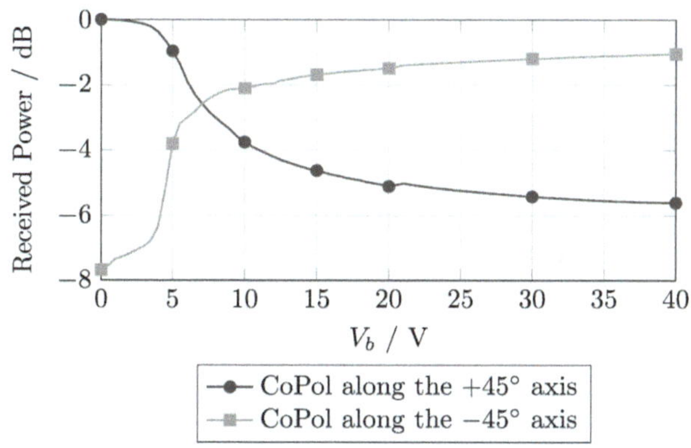

Fig. 5.20 Normalized received power in broadside over the biasing voltage V_b. The co- and cross-polarizations are defined with respect to +45°

The antenna gain is measured at 13.75 GHz by using the gain comparison technique [Bal05]. These results are given in Fig. 5.21 only for 0, 7.5 and 40 V biasing states, accordingly. According to ADS simulations, the antenna gain is 8.23 dB at the broadside when the relative permittivity of the LC layer is equal to 2.46. The simulated gain matches to the corresponding measured one that is 7.9 dB in Fig. 5.21a. The simulation results also show that the antenna features a directivity of 10.7 dB with an efficiency of 57 %. The difference between the simulated directivity and the gain originates from the insertion loss of the feeding network, which is calculated as 2.47 dB. In the circular polarization state, the gain is measured as 5 dB at the broadside as shown in Fig. 5.21b. This is an expected value because it is almost 3 dB less than the one measured in Fig. 5.21a. When the antenna operates at −45° linear polarization state, its gain is measured as 7 dB at the broadside as shown in Fig. 5.21c. These results show that the gain and the cross-polarization level, which are measured

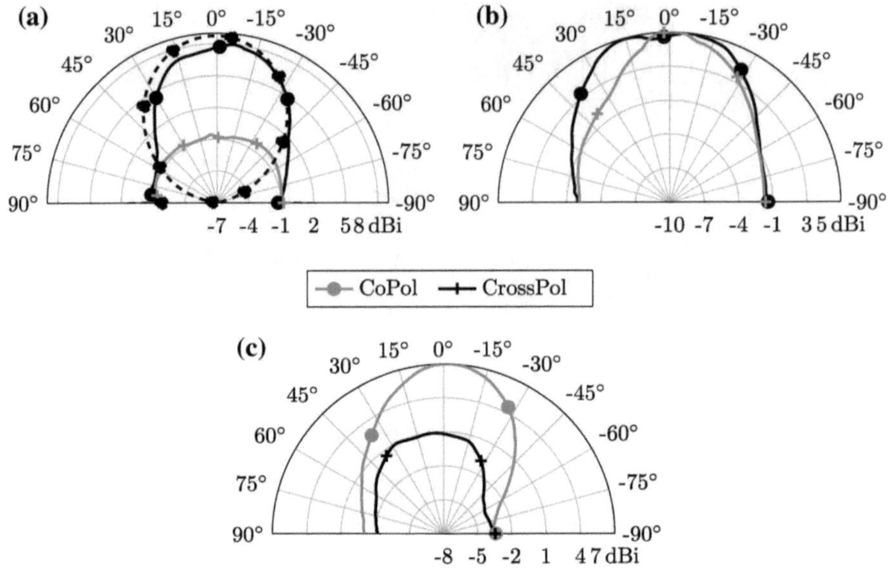

Fig. 5.21 **a** Measured gain pattern of the polarization agile antenna when the biasing voltage $V_b = 0\,\text{V}$. The antenna operates at $+45°$ linear polarization. The *solid* and *dashed lines* represent the measured and simulated data, respectively. The gain pattern of the antenna **b** when the biasing voltage $V_b = 7.5\,\text{V}$, where the antenna operates in the circular polarization, and **c** when the biasing voltage $V_b = 40\,\text{V}$, where the antenna operates in the $-45°$ linear polarization

at the $-45°$ linear polarization state, are 1 and 3 dB less, respectively, compared to those measured at the $+45°$ linear polarization state. The reasons why the gain and the cross polarization level are different for two orthogonal polarization states are discussed in the subsequent paragraphs.

The differential phase shift provided by the RF-feeding networks is the key parameter to obtain the described polarization agility. This can be measured on the antenna system by using a near field scanner. Even then, $\Delta\phi_\beta$ can be measured by realizing the RF-feeding network as a single device to characterize it separately from the antenna. As shown in Fig. 5.22, the RF-feeding networks are modeled with a two-port IMSL variable, which is 37 mm long and includes an IDC. Unlike the RF-feeding networks, this device includes additionally the coplanar waveguide to microstrip line transitions. These are required for the on-wafer measurement probes and their effect can be excluded by doing post-processing.

Measured transmission coefficients of the delay line are plotted in Fig. 5.23 for different bias voltages. The insertion losses are about 3.5 dB at 13.75 GHz. When the effects of the transitions are excluded, the insertion losses are calculated as 2.5 dB. This value matches quite well to the simulated result, which has been mentioned before. The $\Delta\phi_\beta$ versus applied bias voltage characteristic of the IMSL are given in Fig. 5.23 at 13.75 GHz. The device provides a $\Delta\phi_\beta$ of 90° when the bias voltage is about 7.5 V. This confirms the results given in Fig. 5.20, where the antenna operates

Fig. 5.22 A picture of the variable delay line which models the V and H RF-feeding networks. The overall length of the line is 37 mm. It is a 50 Ω transmission line, whose signal width is 0.2 mm

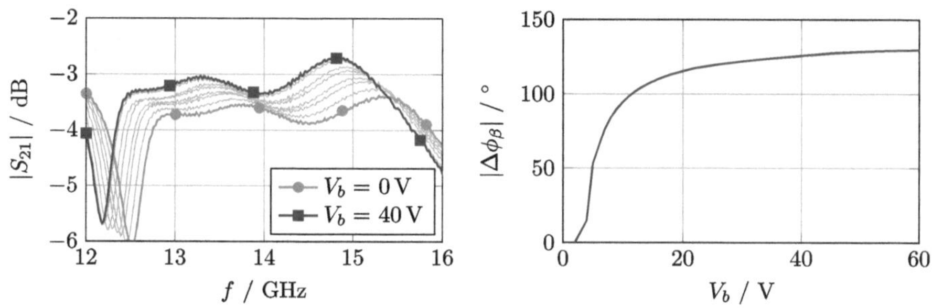

Fig. 5.23 *Left* Magnitudes of the transmission coefficients of the IMSL for different applied voltages V_b. *Right* The differential phase shift $\Delta\phi_\beta$ of the IMSL versus an applied voltage V_b at 13.75 GHz

in the circular polarization if $V_b = 7.5\,\text{V}$. When the applied voltage is increased to 40 V, the $\Delta\phi_{\beta,\max}$ saturates to 130°, which supposed to be 180° to achieve a linear polarization (along the −45° axis). Because of the lack of the desired differential phase shift, the antenna operates at an elliptical polarization state when $V_b = 40\,\text{V}$. Therefore, the measured gain and cross-polarization level are less compared to those at $V_b = 0\,\text{V}$.

Furthermore, the cross polarization levels of the two orthogonal linear polarizations are measured 10 dB less compared to full-wave simulations of the antenna. It is determined that the resistivity of the Cr made bias lines is not sufficiently high and the RF signal, which propagates along the feeding network, leaks through the bias lines. This has been proven by fabrication of a new antenna array, which is shown in Fig. 5.24. This antenna, namely the reference antenna, does not include any bias lines and therefore, it is not polarization agile. The aim is to compare the radiation patterns of the polarization agile antenna and the reference antenna only at the zero biasing case. As shown in Fig. 5.25, the CrossPol level of the reference antenna is higher than that of the polarization agile antenna given in Fig. 5.21a. Particularly, the CrossPol level is increased from 9 to 15 dB at the broadside direction when the

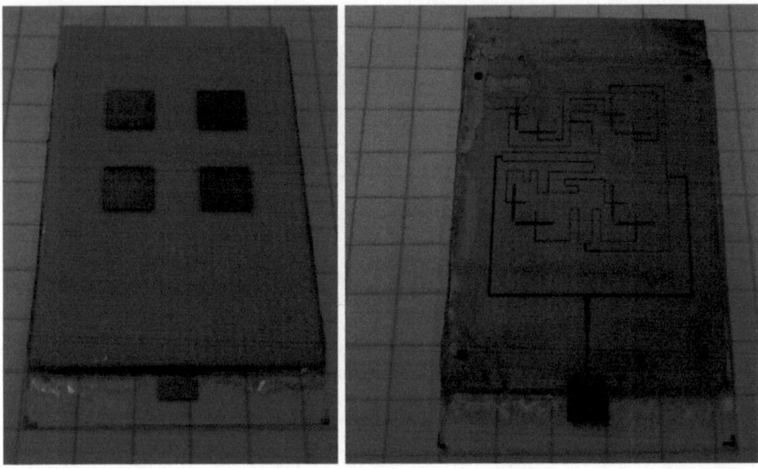

Fig. 5.24 *Left* Pictures of the reference antenna form the *top view*. *Right* From the *bottom view*

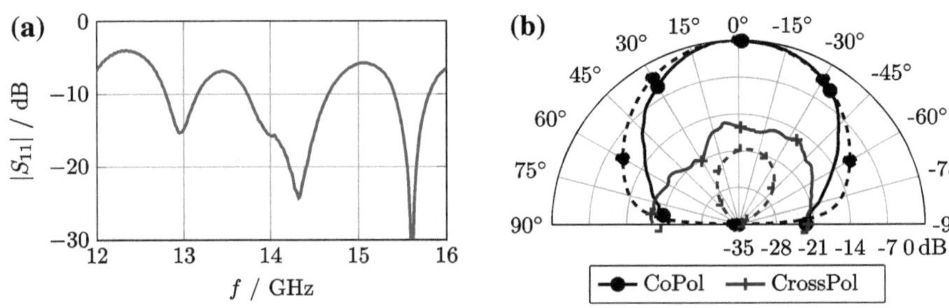

Fig. 5.25 **a** Measured reflection coefficients and **b** measured far-field power pattern of the reference antenna. The antenna operates at +45° linear polarization. The *solid* and *dashed lines* represent the measured and simulated data, respectively

bias lines omitted. Therefore, a low conductive indium tin oxide (ITO) should be preferred for a next fabrication of a polarization agile antenna.

5.3 Summary

This chapter presented two types of polarization agile antennas. Both antennas included a dual-fed radiating element, which could be either a single element or an array, connected to the tunable component. The tunable components were the tunable coupled line and the variable delay line for the first and the second antenna, respectively.

In the first antenna, the tunable coupled line was loaded with shunt LC varactors, which controlled the magnitudes of the two output signal. If the signals were equal in magnitude with a 90° differential phase shift, the antenna operated in the circular polarization state. If the magnitude of the signal is maximized only at one of the output ports, the antenna operated in the linear polarization mode. The tunability of the LC material and that of the LC varactor are less than 0.25. This is not sufficient to obtain the required coupled line characteristics. Hence, methods reducing the required varactor tunability were proposed such as cascading several unit elements or utilizing the back tuning concept. An antenna demonstrator was fabricated and measured to verify the results. The antenna operated at 13.75 GHz with a return loss of better than 10 dB. The measurement results confirm that the antenna polarization state can be changed continuously between linear and circular polarizations.

In the second antenna, the polarization state was tuned by controlling the differential phase shift between $\Delta\phi_\beta$ the two orthogonal antenna feedings. In this manner, the RF-feeding networks were implemented in inverted microstrip line topology to obtain the desired $\Delta\phi_\beta$. The measurement results showed that the antenna operated at 13.75 GHz with a return loss of greater than 16 dB. The polarization state of the antenna was controlled continuously between the dual linear and dual circular polarizations by applying a bias voltage in the range of 0–40 V. According to the measurements results, there were slight discrepancies between the two orthogonal linear polarizations. These were investigated and explained in details by further realizing of corresponding structures.

References

[ARJ75] S. Akhtarzad, T.R. Rowbotham, P.B. Johns, The design of coupled microstrip lines. IEEE Trans. Microw. Theory Tech. **23**(6), 486–492 (1975)

[Bal05] C.A. Balanis, *Antenna Theory: Analysis and Design*, 3rd edn. (Wiley, New York, 2005)

[Cor] Rogers Corporation. RO4000^{TM} series high frequency circuit materials

[Fer+09] F. Ferrero, C. Luxey, R. Staraj, G. Jacquemod, M. Yedlin, V. Fusco, A novel quad-polarization agile patch antenna. IEEE Trans. Antennas Propag. **57**(5), 1563–1567 (2009)

[GSZ06] S. Gao, A. Sambell, S.S. Zhong, Polarization-agile antennas. IEEE Antennas Propag. Mag. **48**(3), 28–37 (2006)

[KJ84] M. Kirschning, R.H. Jansen, Accurate wide-range design equations for the frequency-dependent characteristic of parallel coupled microstrip lines. IEEE Trans. Microw. Theory Tech. **32**(1), 83–90 (1984)

[MMH92] F. Masot, F. Medina, M. Horno, Theoretical and experimental study of modified coupled strip coupler. Electron. Lett. **28**(4), 347–348 (1992)

[Mue07] S. Mueller, Grundlegende Untersuchungen steuerbarer passiver Flüssigkristall-Komponenten für die Mikrowellentechnik. PhD thesis, Technische Universitaet Darmstadt, 2007

[Poz11] D.M. Pozar, *Microwave Engineering*, 4th edn. (Wiley, New York, 2011)

[Saz+11b] M. Sazegar, A. Mehmood, Y. Zheng, H. Maune, R. Jakoby, Low-cost phased-array antenna using compact tunable phase shifters based on ferroelectric ceramics, in *IEEE Transactions on Microwave Theory and Techniques*, vol. 59, pp. 1265–1273, 2011

[SCHd] SCHOTT. Synthetic fused silica. http://www.schott.com/korea/korean/download/datasheet_fused_silica_.pdf

[SS74] B. Sheleg, B.E. Spielman, Broad-band directional couplers using microstrip with dielectric overlays. IEEE Trans. Microw. Theory Tech. **22**(12), 1216–1220 (1974)

[ZJ69] G.I. Zysman, A.K. Johnson, Coupled transmission line networks in an inhomogeneous dielectric medium. IEEE Trans. Microw. Theory Tech. **17**(10), 753–759 (1969)

Chapter 6
Conclusion and Outlook

This work presented electronic beam steering and polarization agile planar antennas implemented in liquid crystal (LC) technology. All required details starting from an antenna design until its measurement have been described systematically. This work can be listed as the precise broadband characterizations of nematic LCs and dielectric substrates, equivalent circuit and layout designs of tunable devices and finally, realizations and measurements of these devices.

All LC materials used in this work were provided by Merck KGaA, Darmstadt together with the material parameters extracted at 19 GHz and at the room temperature. The material parameters are correlated with the operating frequency and temperature, and therefore, variations in either frequency or temperature has impact on the antenna performance. This work offers a temperature-controlled artificial coaxial line, enabling the characterization of the LC materials over a wide range of both frequency and temperature. These ranges are from 3.5 to 30 GHz and from 10 °C to 60 °C, respectively. The measurement results have shown the well-know feature of LCs [Mue07], that is both, the material tunability and efficiency increase with increasing frequency. It has been also shown that microwave optimized LCs can provide desired performances over an extended operating temperature range compared to LCs used in displays such as K15 (5CB). However, unlike the frequency dependency, both, the material tunability and efficiency decrease with increasing temperature.

Another aspect of this work is characterization of several glass substrates for microwave frequencies to figure out low-loss and, especially, cost-effective substrates. The glass substrates were first characterized at 20 GHz by using the split-cylinder-cavity method. Then, the ones with promising results were patterened with coplanar waveguides for further broadband characterization. As a results, three low-loss glass substrates with $\tan\delta < 0.009$ were determined, are at least 10 times cheaper than the fused silica, which is a typical low-loss glass substrate ($\tan\delta < 0.0015$) used in the microwave applications.

Concerning electronic beam steering planar antennas, this work advances the state-of-the-art for the LC based antennas by enabling 2D beam steering. This was

O. H. Karabey, *Electronic Beam Steering and Polarization Agile Planar Antennas in Liquid Crystal Technology*, Springer Theses, DOI: 10.1007/978-3-319-01424-1_6,

achieved by implementing the LC delay lines in a more compact way compared to, for example, [KFN02], by means of a new geometry. They were incorporated with the biasing network on the same wafer as the antenna element. Before coming up with this solution, several LC delay lines were studied. These devices were implemented in by using planar transmission lines such as coplanar waveguide, slotline and microtrip lines. Compared to a hollow waveguide, these lines feature higher losses. However, those transmission lines were chosen to realize low-profile antennas. The delay lines studied in this work can be categorized into two types, namely transmission line type and loaded line type delay lines. In the first type of devices, the LC material is utilized as tunable dielectric of a chosen transmission line and in the second one, a chosen transmission line is periodically loaded with LC varactors.

The delay lines were first designed in equivalent circuit level. At this point, the design procedure was not dedicated either for an operating frequency or for a transmission line topology or even not for a tunable device technology. Indeed, these results can be applicable for any of those given requirements. However, while transforming the results from the equivalent circuit to the layout level, a special attention has been given to optimize device geometries. Hence, the available LC material tunability can be exploited as efficiently as possible by reducing parasitics occurring in the layout. As a result, firstly, transmission line type delay line, namely inverted microstrip line (IMSL), was proposed in this work, which features a high delay line efficiency η_{VDL} that is 90 °/dB at 17.5 GHz. This is obtained by selecting thinnest glass substrate with a relative permittivity as close as possible to that of the LC material. Hence, the fringing fields are minimized and almost all the electrical fields are propagated through the LC material. Secondly, a loaded coplanar waveguide delay line was realized by using comparatively lossy whereas cost-effective glass substrate unlike to [Goe09]. Additional material losses were compensated by increasing the quality factor of the varactors with a trapezoidal geometry, which reduces the parasitic inductances. Measurement result showed that the proposed varactor features a quality factor of higher than 120 at 20 GHz. Similar to [Goe09], the delay line efficiency was obtained around 60 °/dB at 20 GHz even with the lossy glass material. Thirdly, a loaded slotline device is realized, which enables to implement LC varactors in parallel plate capacitor topology, leading to high tuning efficiency. The delay line efficiency was measured as 42.5 °/dB at 18 GHz. This is still a high value compared to ones studied by using RF MEMS [CVH08] or ferroelectrics [Mir+08], which provides 28 °/dB at 10 GHz and 25 °/dB at 30 GHz, respectively. Finally, a loaded microstrip line topology is proposed to exploit the advantage of low insertion losses featured by the microstrip line technology. This device was implemented by loading the microstrip line with open split ring resonators (OSRR). It was shown that the quality factor of OSRR was less than 20 at 16 GHz, leading to a delay line efficiency of 25 °/dB at the same frequency.

Both loaded slotline and microstrip line topologies were studied aiming to reduce device size and balance the efficiency and response times compared to the inverted microstrip line topology. Although low delay line efficiencies are measured for both, which are 42.5 °/dB at 18 GHz and 25 °/dB at 16 GHz, respectively, the knowledge and experience gained during these studies led to a new and promising approach

Table 6.1 Comparison of electronic beam steering planar phased arrays

	Technology	f_0 (GHz)	Element number	Return loss (dB)	Beam steering	G_{max} (dBi)	SLL_{max} (dB)
[Top+08]	RF MEMS	14.6	1 × 4	>30	1D, 0° to 14o	-	−12
[KPP08]	RF MEMS	14	2 × 2	>12	1D, −8° to 4°	7.75	−8
[Gau+09]	RF MEMS	35	4 × 4	>30	1D, −15° to 15°	8	−8
[Bag+10]	RF MEMS Semicon.	11.75	4 × 4	>12	1D, 0° to 60°	18	−12
[Saz+11b]	BST	10	3 × 4	>17	1D, −50° to 50°	-	−12
[SHK09]	LC-LTCC	30	4 × 4	>25 (Sim.)	1D, −30° to 30°	-	−13
This work	LC	17.5	2 × 2	> 15	2D, −25° to 25°	5.9	−8

of a serially loaded microstrip delay line. This new topology is subject to a current patent application, and therefore, it is not described herein. It combines high delay line efficiency of around 60 °/dB at 20 GHz with low response times of less than 40 ms and moderate physical length of less than $2.33\lambda_0$ to provide 360 ° differential phase shift. Nevertheless, as being the major goal of this work in a limited time, a 2D electronic beam steering antenna is implemented by using the IMSL topology as proof-of-concept, because of its high efficiency. The antenna consists of 2 × 2 microstrip patch array, variable delay lines, dc-blocking structures, RF feeding, and biasing networks. All these components were implemented into three dielectric layers with an overall thickness of 1.5 mm. The key parameters obtained in this work are compared to those presented in the literature in Table 6.1. Here, electronic beam steering and planar phased arrays, which are designed to operate at higher than 10 GHz, are chosen. The antenna proposed in this work provides 2D beam steering capability since each delay line can be biased individually. The antenna presented in [Bag+10] is also designed for a 2D beam steering application where the main beam is steered by using hybrid tracking technique. Therefore, the antenna profile is increased considerably. The antennas proposed in this work and in [Saz+11b] are continuously steerable because of the chosen technologies. The others can point the main beam direction towards predefined beam angles, which are defined by the resolutions of their tunable components. The gain measured in this work can be compared to the one given in [KPP08] since both have 2 × 2 radiating elements and operate at close frequencies. The LC based antenna provides 1.8 °/dB less gain than the one reported in [KPP08]. In fact, the second antenna can steer the main beam from −88 °. This requires a short delay line with a $\Delta\phi_{\beta,\max}$ of 33 °. In this case, the insertion losses caused by the delay lines are less and so, the antenna gain increases.

Beside the electronic beam steering antenna, two types of polarization agile antennas were studied in LC technology. To the author's knowledge, these kind of antennas were studied for the first time in LC technology. The first antenna includes a tunable coupled line, which is connected to a dual-fed microstrip patch antenna. The polarization state of the antenna is controlled by tuning the coupling coefficient of the tunable coupled line. This architecture has been proposed in [Fer+09] for an S-band application by using semiconductor varactors having a varactor tunability of more than 0.8. In this work, however, the focus was to obtain desired coupled line functionalities by having an LC varactor tunability of 0.2 at maximum. The demonstrator verified successfully the continuous polarization agility between a linear and right-handed circular polarizations at 13.75 GHz. Empirical studies showed that a polarization agility between dual linear and dual circular polarizations can also be achieved if the varactor tunability is about 0.48. The second antenna includes dual-fed microstrip patch antennas in a 2×2 array configuration. The horizontal (H) and vertical (V) feedings of the dual-fed antennas are connected to each other through two separate RF-feeding networks. The polarization state of the antenna is controlled by introducing a differential phase shift between the H and V feedings. Owing to the LC technology, the RF-feeding networks were implemented in the IMSL topology. Thus, a desired differential phase shift was provided by the RF-feeding networks without using any additional delay line. The measurement results confirmed the continuous polarization agility between dual linear and dual circular polarizations. The underlying results are listed in Table 6.2 and compared with the published results.

Most of the polarization agile planar antennas have been realized by using switches, enabling only to switch between possible antenna polarization states. In contrary, the antennas realized with varactors as in this work provide additional polarization matching because of continuous tuning of varactors. Another common point of these previous studies is that the antennas mainly operate around L- and S-band. However, the antennas proposed in this work and some of the RF MEMS based antennas operate at higher microwave frequencies ($f > 10$ GHz). Indeed, the LC based antennas can be scaled to operate at Ku-/Ka-bands and at even higher, i.e. W-band, frequencies because of low dielectric losses.

Both the electronic beam steering phased array and polarization agile antennas were implemented by adapting the well-established LC display technology. They were fabricated by using in-house facilities whereas they can be easily realized utilizing automated manufacturing techniques. This enables a low-cost fabrication of larger arrays even for a low volume production. Owing to low-profile and planarity, the proposed antenna concepts are convenient for applications, where antennas are aimed to be mounted close or even integrated into a structure, e.g., in a rooftop or body of a vehicle for terrestrial and satellite based services.

The realized electronic beam steering phased array antenna is limited to operate in portable applications. For a mobile application, on the other hand, a loaded delay line topology with low response time is required. In the lights of presented results, a new serially loaded microstrip line topology is proposed. This comprises a high delay line efficiency of around 60 °/dB at 20 GHz and low response time of less than 40 ms.

Table 6.2 Comparison of polarization agile antennas. AR stands for axial ratio

	Technology, device	f_0 (GHz)	Polarization agility		Crosspol (dB)	AR (dB)
[Ems]	Ferrite, Phase Shifter	X-/Ka-Band	Dual Dual Lin. ↔ Circ.	Switched	-	-
[Wan+11]	RF MEMS, Switch	14 35	Dual Linear	Switched	−15	-
[Jun+12]	RF MEMS, Switch	20	Lin. ↔ Circ.	Switched	−12	0
[BDF11]	RF MEMS, Switch	4-7	Dual Dual Lin. ↔ Circ.	Switched	-	-
[SCK02]	RF MEMS, Actuators	26.7	Dual Single Lin. ↔ Circ.	Switched	-	2
[WR06]	Semiconductor, PIN Diode	2.45	Dual Circular	Switched	-	1.5
[Sch+81]	Semiconductor, PIN Diode	1.47	Dual Dual Lin. ↔ Circ.	Switched	< -30	1
[SF02]	Semiconductor, PIN Diode	1	Lin. ↔ Circ.	Switched	−18	2.5
[HD94]	Semiconductor, PIN Diode	1.5	Dual Dual Lin. ↔ Circ.	Switched	−20	1
[Fer+09]	Semiconductor, Varactor	3.5	Dual Dual Lin. ↔ Circ.	Continous	−21	0.8
[PCD10]	Semiconductor, Varactor	0.88	Lin. ↔ Circ.	Continous	-	2.2
This work	LC, varactor	13.75	Lin. ↔ Circ.	Continous	−12	-
This work	LC, Delay line	13.75	Dual Dual Lin. ↔ Circ.	Continous	−9	-

This new and promising topology requires further investigations, and in fact, it is subject to one of the an on-going industrial projects.

In this work, the bias lines were made of chromium or nickel/chromium layers, which were used as adhesive layers between a glass substrate an RF electrode such as gold. It has been experimentally shown that the resistivity of the bias lines were not sufficiently high and so that, they have impact on the RF circuitry. For instance, the return loss of the electronic beam steering antenna was increased from −30 to −20 dB and similarly, the cross polarization level of the polarization agile antenna was increased to −15 to −9 dB. Therefore, new adhesive layers should be investigated, leading to high resistive bias lines.

Finally, both beam steering and polarization agility can be combined in one antenna. The challenge is to reduce the delay lines size further, because two times more delay lines are required since the antenna features both reconfigurability simultaneously. In this case, the results of this work form a basis for optimization of the delay lines. Such an antenna can be operated at 60 GHz for indoor applications such as connecting to wireless local access networks (WLANs). As having a low profile, the antenna can be integrated into a mobile terminal for some aesthetic reasons, i.e. into the cover of a notebook.

References

[Mue07] S. Mueller, Grundlegende Untersuchungen steuerbarer passiver Flnssigkristall-Komponenten fnr die Mikrowellentechnik. PhD thesis, Technische Universitaet Darmstadt, 2007

[KFN02] T. Kuki, H. Fujikake, T. Nomoto, Microwave variable delay line using dual-frequency switching-mode liquid crystal. IEEE Trans. Microw. Theory Tech. **50**, 2604–2609 (2002)

[Goe09] F. Goelden, Liquid Crystal Based Microwave Components with Fast Response Times: Material, Technology, Power Handling Capability. PhD thesis, Technische Universitaet Darmstadt, 2009

[CVH08] K.V. Caekenberghe, T. Vaha-Heikkila, An analog rf mems slotline true-time-delay phase shifter. IEEE Trans. Microw. Theory Tech. **56**(9), 2151–2159 (2008)

[Mir+08] I.G. Mironenko, A.A. Semenov, A.A. Ivanov, P.J. Beljavski, C.-Y. Kang, S.F. Karmanenko, Multislot transmission lines and microwave phase shifters based on (ba, sr)tio3 films, in *Integrated Ferroelectrics: an, International Journal*, vol. 97.1, 2008, pp. 58–66. ISSN: 0018–9480

[Bag+10] L. Baggen, S. Vaccaro, D. Llorens del Rio, G. Langgartner, Compact phased arrays for mobile terminals, in *Proceedings of the International Semiconductor Conference (CAS)*, vol. 01, 2010, pp. 3–9

[Saz+11b] M. Sazegar, Z. Yuliang , H. Maune, C. Damm, Z. Xianghui, J. Binder, R. Jakoby, Low-cost phased-array antenna using compact tunable phase shifters based on ferroelectric ceramics. IEEE Trans. Microw. Theory Tech. **59**, 1265–1273 (2011)

[KPP08] N. Kingsley, G.E. Ponchak, J. Papapolymerou, Reconfigurable rf mems phased array antenna integrated within a liquid crystal polymer (lcp) system-on-package. IEEE Trans. Antennas Propag. **56**(1), 108–118 (2008)

[Top+08] K. Topalli, O.A. Civi, S. Demir, S. Koc, T. Akin, A monolithic phased array using 3-bit distributed rf mems phase shifters. IEEE Trans. Microw. Theory Tech. **56**(2), 270–277 (2008)

[Gau+09] W. Gautier, V. Ziegler, A. Stehle, B. Schoenlinner, U. Prechtel, W. Menzel, Rf-mems phased array antenna on low-loss ltcc substrate for ka-band data link. in *Proceedings of the European Microwave Conference EuMC 2009*, 2009, pp. 914–917

[SHK09] B. Sanadgol, S. Holzwarth, J. Kassner. 30 ghz liquid crystal phased array, in *Loughborough Antennas and Propagation Conference LAPC 2009*, 2009

[Fer+09] F. Ferrero, C. Luxey, R. Staraj, G. Jacquemod, M. Yedlin, V. Fusco, A novel quad-polarization agile patch antenna. IEEE Trans. Antennas Propag. **57**(5), 1563–1567 (2009)

[Ems] Application note pol-1: Phased array polarization switches. Technical report, EMS Technologies, Inc. Space and Technology - Atlanta, 2003

[Wan+11] G. Wang, R. Bairavasubramanian, B. Pan, J. Papapolymerou. Radiofrequency mems-enabled polarisation reconfigurable antenna arrays on multilayer liquid crystal polymer, in *Microwaves, Antennas Propagation, IET*, vol. 5.13, pp. 1594–1599. ISSN: 1751-8725

[Jun+12] T.J. Jung, I.-J. Hyeon, C.-W. Baek, S. Lim, Circular/linear polarization reconfigurable antenna on simplified rf-mems packaging platform in k-band, in *Antennas and Propagation, IEEE Transactions on, 60.11*, 2012, pp. 5039–5045. ISSN: 0018–926X

[BDF11] A.G. Besoli and F. De Flaviis, A multifunctional reconfigurable pixeled antenna using mems technology on printed circuit board, in *Antennas and Propagation, IEEE Transactions on, 59.12*, 2011, pp. 4413–4424. ISSN: 0018–926X

[SCK02] R.N. Simons, D. Chun, L.P.B. Katehi, Polarization reconfigurable patch antenna using microelectromechanical systems (mems) actuators, in *Antennas and Propagation Society International Symposium, 2002. IEEE*, vol. 2, 2002, pp. 6–9

[WR06] J.-F. Wu, J.-S. Row, Broadband circularly-polarised microstrip antenna with switchable polarisation sense. Electron. Lett. **42**(24), 1374–1375 (2006)

[Sch+81] D. Schaubert, F. Farrar, A. Sindoris, S. Hayes, Microstrip antennas with frequency agility and polarization diversity. IEEE Trans. Antennas Propag. **29**(1), 118–123 (1981)

[SF02] H. Scott, V.F. Fusco, Polarization agile circular wire loop antenna. IEEE Antennas Wirel. Propag. Lett. **1**(1), 64–66 (2002)

[HD94] P.M. Haskins, J.S. Dahele, Varactor-diode loaded passive polarisation-agile patch antenna. Electron. Lett. **30**(13), 1074–1075 (1994)

[PCD10] D. Piazza, M. Capacchione, M. D'Amico, Crlh leaky wave antenna with tunable polarization, in *proceedings of the fourth european conference on antennas and propagation (EuCAP)*, 2010, pp. 1–5

Appendix A

A.1 Technology

A.1.1 Fabrication Process Parameters

- **Thermally Evaporated Seed Layer**

Nickel-Chromium (Ni-Cr)	: Thickness, 15 nm
Gold (Au)	: Thickness, 60 nm

- **Photolithography** (Table A.1)

Hard bake	: 15 min at 120 °C
O_2 etching	: Flow rate of 100 sccm at 20 kPa 20 s at 200 W

- **Gold Plating**

Plating bath	: Puramet 402: 10 g/L
Temperature	: 37 °C
Time	: $\approx$1 h for 1 μm thickness when the current density is 0.5 mA/cm^2

O. H. Karabey, *Electronic Beam Steering and Polarization Agile Planar Antennas in Liquid Crystal Technology*, Springer Theses, DOI: 10.1007/978-3-319-01424-1,

Table A.1 Photolithography process parameters

Required electrode thickness (after electroplating)	2 μm	5 μm
Photoresist [Mic]	AZ4533	AZ4562
Spin coating	40 s at 4000 rpm	40 s at 3000 rpm
Pre exposure bake	3 min at 90 °C	6 min at 90 °C
UV-exposure time ($6\,mW/cm^2$)	35 s	75 s
Developer [Mic]	AZ400K:H_2O (1:4)	AZ400K:H_2O (1:4)
Development time	40 s at 23 °C	70 s at 23 °C

• **Photoresist Stripping**

Hot acetone : 5 min at 60 °C
Acetone injection
Isopropyl alcohol injection
O_2 etching : Flow rate of 100 sccm at 20 kPa
5 min at 200 W

• **Wet Etching**

Nickel-Chromium etching
Etchant
$NH_4[Ce(NO_3)_6]$:HNO_3:H_2O
(41.15 g:22.5 ml:250 ml)
Temperature 23 °C
Time 3 s

Gold etching
Etchant
I_2:KI:H_2O (5 g:20 g:200 ml)
(5 g:20 g:200 ml)
Temperature 23 °C
Time 15 s

A.1.2 Alignment Layer Process Parameters

- **Constituents of the alignment Layer [Goe09]**

Nylon6	$[NH\text{-}(CH_2)_5\text{-}CO]_n$	1.5 g
Trichlorethanol	$C_2H_3Cl_3O$	100 g

- **Spin Coating Parameters**

The alignment layer orients the liquid crystal molecules at the surfaces parallel to the surfaces in case of no external field. The spin coating parameters of the alignment layer are given as 6000 rpm at 30 s in [Goe09]. These parameters are essential for an optimized device performance and therefore, several experiments have been performed in the frame of this work. For a starting point, three gold electrodes are coated with an alignment layer by using different process parameters. The values given in [Goe09] are used as a reference and two parameters are tested as well. These new parameters are 3000 rpm and 2000 rpm at 30 s. The covered surfaces are rubbed mechanically by using a rubbing machine [Goe09]. An atomic force microscope is utilized to observe the thickness and the uniformity of the grooves. As can be seen in Fig. A.1, the grooves are more uniform when the coating is performed slowly. The measured surface roughnesses and groove thicknesses are given in Table A.2.

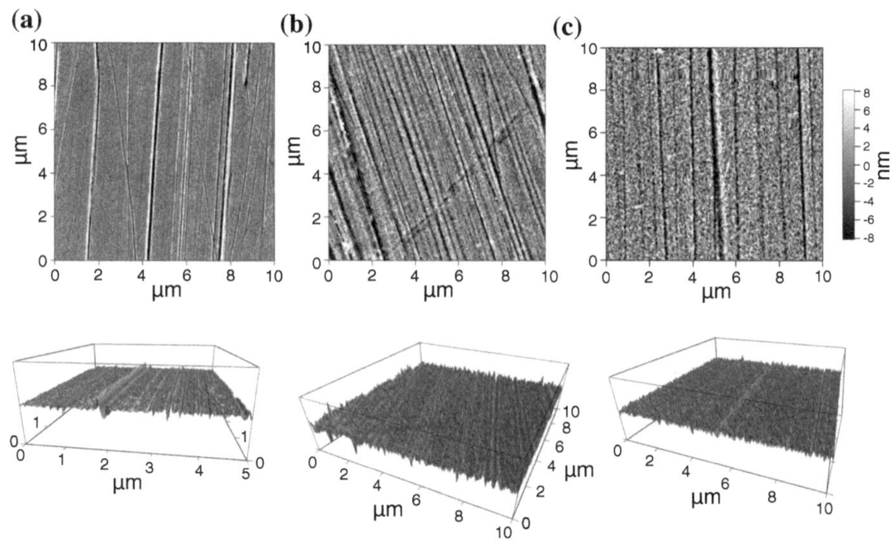

Fig. A.1 Pictures of the surfaces that are coated with the alignment layer. The top views are given on the *top* and the perspective views are given on the *bottom*. The surfaces are scanned by using an atomic force microscope. **a** The surface is spin coated with 6000 rpm at 30 s. **b** 3000 rpm at 30 s and **c** 2000 rpm at 30 s

Table A.2 Measured surface scanning results of alignment layer coated substrates

Spin coating parameter (rpm)	Surface roughness (nm)	Groove thickness (nm)
6000	3.572	20
3000	3.622	30
2000	3.317	43

As the coating speed reduces, the surface flattens and the groove thickness increases. Hence, the loss caused by the surface roughness can be reduced. At the same time the response time of the LC devices can be reduced when the molecules are aligned by means of the surface anchoring.

- **Curing Process of the alignment Layer**

Hotplate	10 min at 90 °C
Oven	23–90 °C in 30 min
	60 min at 90 °C
	90–180 °C in 30 min
	60 min at 180 °C

- **Effects of alignment Layer Process Parameters on Variable Delay Line Performance**

The effects of the alignment layer coating parameters on the variable delay line are presented here. The both signal and ground electrodes of four identical inverted microstrip lines are coated with different parameters. There is another reference variable delay line, which has no alignment layer on its electrodes. Other parameters are 1000 rpm, 1500 rpm, 2000 rpm and 6000 rpm at 30 s. Scattering parameters of the delay lines are measured. These results are given in Figs. A.2, A.3 and A.4 for

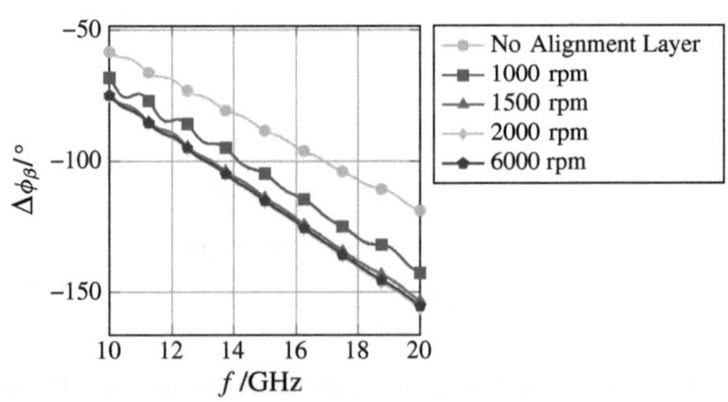

Fig. A.2 Measured differential phase shifts $\delta\phi_\beta$ of variable delay lines for different coating parameters

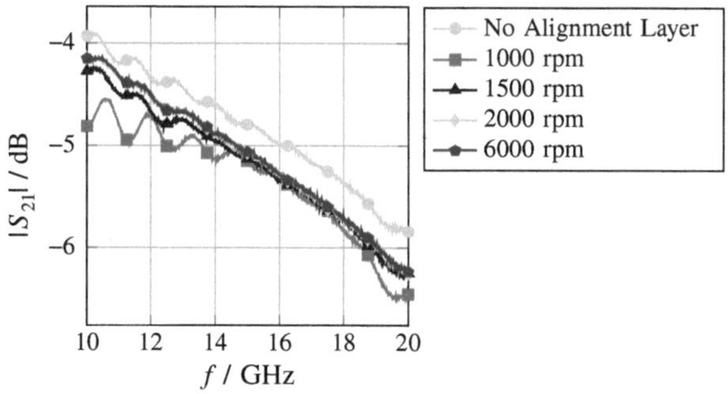

Fig. A.3 Measured magnitudes of transmission coefficients of variable delay lines for different coating parameters

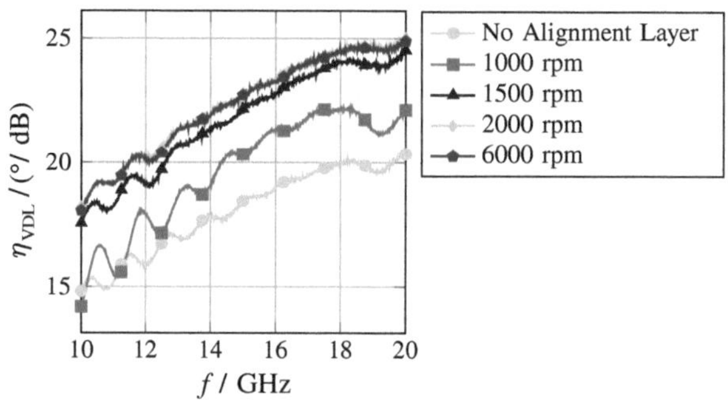

Fig. A.4 Measured variable delay line efficiencies η_{VDL} for different coating parameters

differential phase shift $\Delta\phi_\beta$, for magnitudes of the transmission coefficients $|S_{21}|$ and for variable delay line efficiency η_{VDL}, respectively.

The results are listed in Table A.3 for 18 GHz. In the absence of this layer, the LC molecules can not achieve the desired alignment completely. Hence, the resultant $\Delta\phi_\beta$ is less and t_{off} is approximately 2.5 times more compared to those of the delay lines with the alignment layer. On the other hand, t_{on} is much faster since there is no resisting anchoring force to the applied electrical field. In the presence of the alignment layer, the measured results are close to each other. This means that improvements on the alignment layer do not have significant impact on the device performance because of the other fabrication tolerances. Therefore, no more investigations are carried out. Nevertheless, the coating parameters are specified to be 2000 rpm at 30 s for the devices realized in this work because the maximum η_{VDL} is measured with these parameters.

Table A.3 Measured results for different alignment layer parameters at 23 °C at 18 GHz

Spin coating parameter (30 s)	$\Delta\phi_\beta$ (°)	$\|S_{21}\|$ (dB)	η_{VDL} (°/dB)	t_{on} (s)	t_{off} (s)
No alignment layer	107.18	−5.37	19.97	0.058	157
1000 rpm	129.22	−5.84	22.13	0.236	69.58
1500 rpm	138.24	−5.75	24.05	0.236	63.51
2000 rpm	**140.63**	**−5.72**	**24.56**	**0.232**	**84.37**
6000 rpm	139.72	−5.71	24.48	0.257	70.05

A.2 Formulas for Microwave Characterization

A.2.1 Nicolson-Ross-Weir Method

The complex permittivity and permeability of a material can be calculated by using the analytical formulas derived in Nicolson-Ross-Weir (NWR) method [NR70, Wei74, BJ+93, Che+04]. In this method the reflection coefficient Γ is calculated by

$$\Gamma = X \pm \sqrt{X^2 + 1}, \tag{A.1}$$

where

$$X = \frac{S_{11}^2 - S_{21}^2 + 1}{2S_{11}}. \tag{A.2}$$

directly from the scattering parameters $S_{i,j}$. The correct sign in Eq. (A.1) is chosen by providing $|\Gamma| \geq 1$. The transmission coefficient T is determined by

$$T = \frac{S_{11} + S_{21} - \Gamma}{1 - (S_{11} + S_{21})\Gamma} \tag{A.3}$$

Based on the predetermined Γ and T, the unknown complex relative permeability $\underline{\mu}_r$ and permittivity $\underline{\varepsilon}_r$ can be calculated as

$$\underline{\mu}_r = \frac{(1 + \Gamma)}{(1 - \Gamma) \wedge \sqrt{1/\lambda_0^2 - 1/\lambda_c^2}} \tag{A.4}$$

and

$$\underline{\varepsilon}_r = \frac{\lambda_0^2}{\underline{\mu}_r (1/\lambda_c^2 - 1/\wedge^2)}, \tag{A.5}$$

respectively, with

$$\frac{1}{\wedge^2} = -\left(\frac{1}{2\pi \ell_{\text{physical}}} \ln\left(\frac{1}{T}\right)\right)^2, \tag{A.6}$$

In Eqs. (A.4) and (A.5), λ_0 is the free space wavelength and λ_c is the cutoff wavelength of a chosen transmission line. For instance, $\lambda_c = \infty$ if a coaxial line is used. In Eq. (A.6), ℓ_{physical} is the physical length of the transmission line.

Calculating the natural logarithm of the complex transmission coefficients leads multiple solutions for Eq. (A.6). This problem can be solved by either comparing the calculated and measured group delays [BJ+93] or using the phase unwrapping method proposed in [Che+04].

Extracted Material Parameters

This subsection presents the extracted material parameters of the microwave LC materials as a continuing of Sect. 3.1.3. The results are given when the molecules are aligned parallel (∥) and perpendicular (⊥) with respect to the electric field of the coaxial line. After aligning the molecules, scattering parameters are measured while heating the setup from 10 to 60 °C in 5 °C steps (Figs. A.5, A.6, A.7, A.8, A.9, A.10, A.11, A.12).

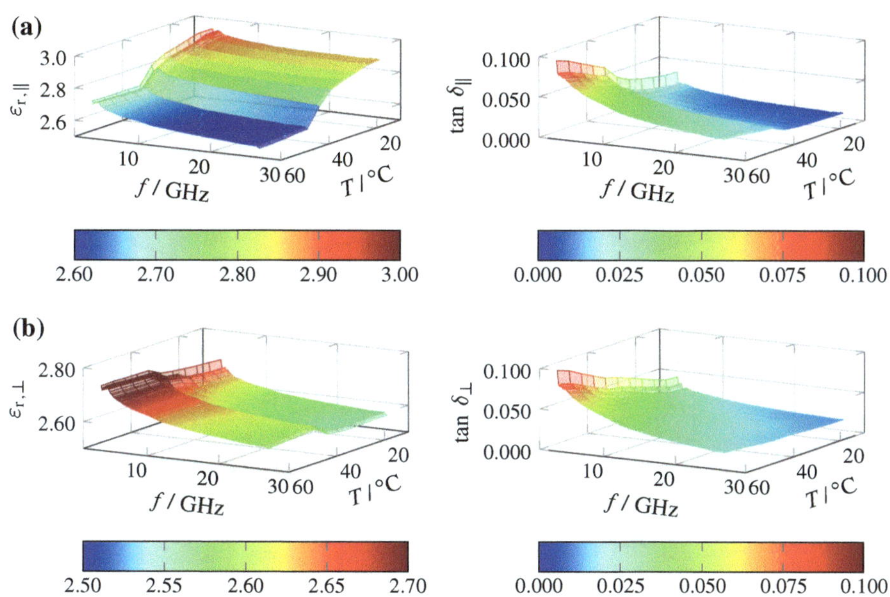

Fig. A.5 Extracted relative permittivity ε_r and loss tangent $\tan\delta$ of K15 versus frequency f and temperature T. **a** The liquid crystal molecules are aligned parallel (∥). **b** The liquid crystal molecules are aligned perpendicular (⊥)

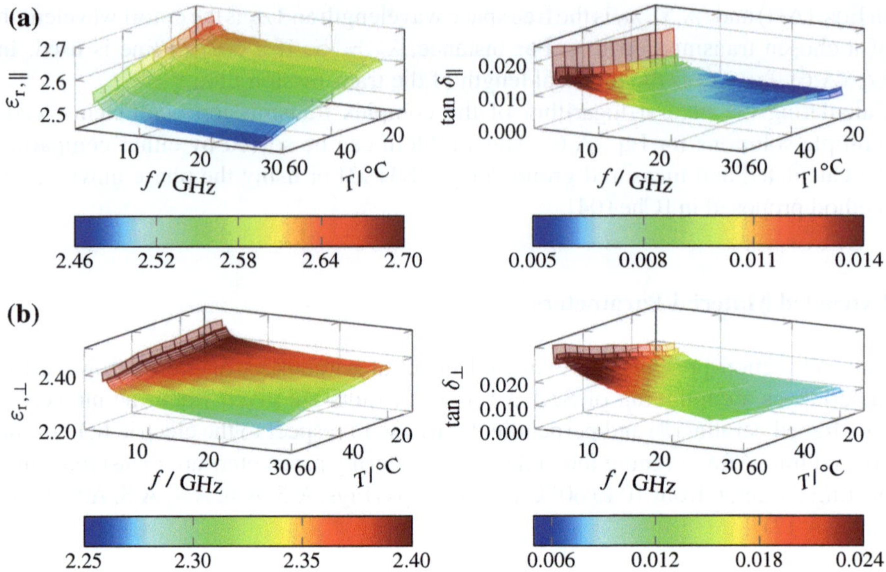

Fig. A.6 Extracted relative permittivity ε_r and loss tangent $\tan\delta$ of TUD-823 versus frequency f and temperature T. **a** The liquid crystal molecules are aligned parallel (‖). **b** The liquid crystal molecules are aligned perpendicular (⊥)

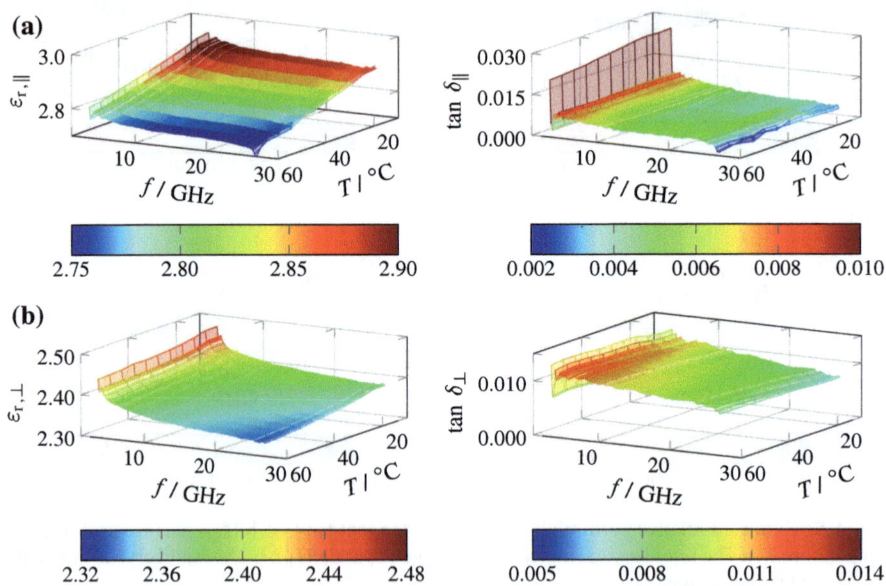

Fig. A.7 Extracted relative permittivity ε_r and loss tangent $\tan\delta$ of TUD-670 versus frequency f and temperature T. **a** The liquid crystal molecules are aligned parallel (‖). **b** The liquid crystal molecules are aligned perpendicular (⊥)

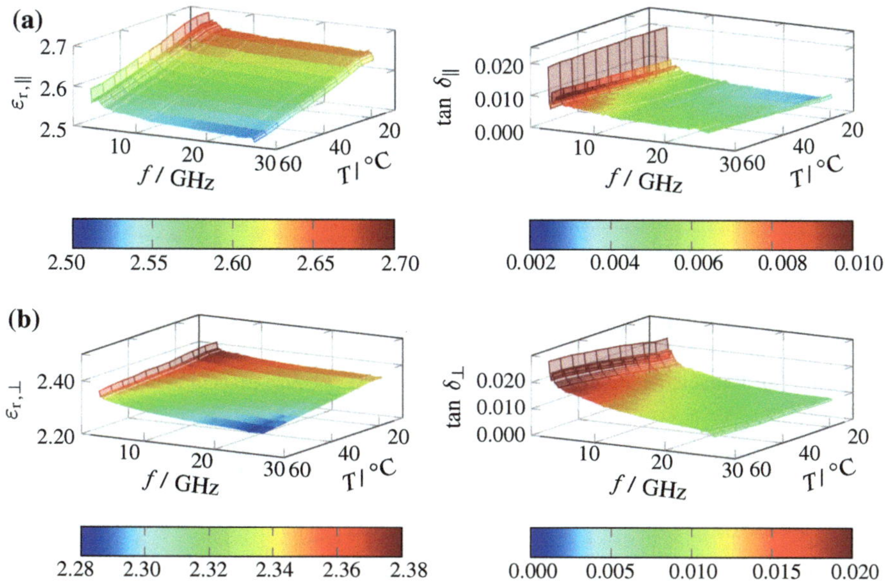

Fig. A.8 Extracted relative permittivity ε_r and loss tangent $\tan\delta$ of TUD-862 versus frequency f and temperature T. **a** The liquid crystal molecules are aligned parallel ($\parallel$). **b** The liquid crystal molecules are aligned perpendicular ($\perp$)

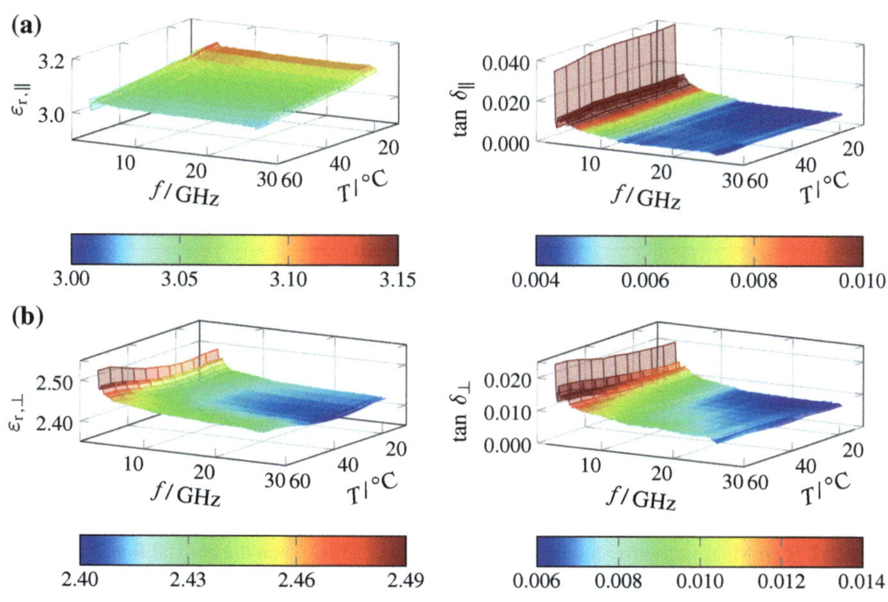

Fig. A.9 Extracted relative permittivity ε_r and loss tangent $\tan\delta$ of TUD-899 versus frequency f and temperature T. **a** The liquid crystal molecules are aligned parallel ($\parallel$). **b** The liquid crystal molecules are aligned perpendicular ($\perp$)

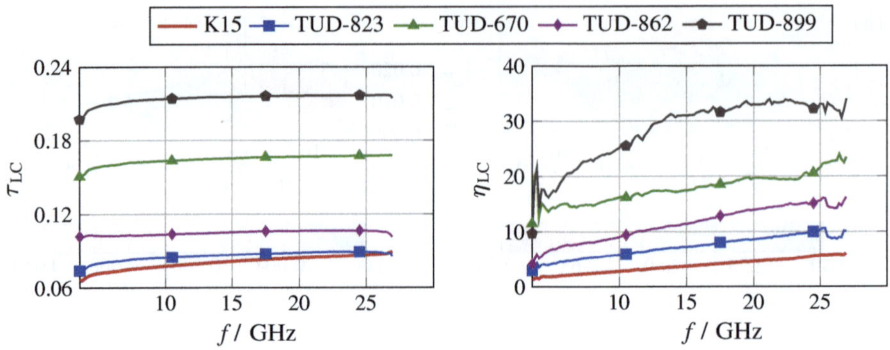

Fig. A.10 *Left* Extracted material tunabilities τ_{LC} at 25 °C. *Right* Extracted material efficiencies η_{LC} at 25 °C

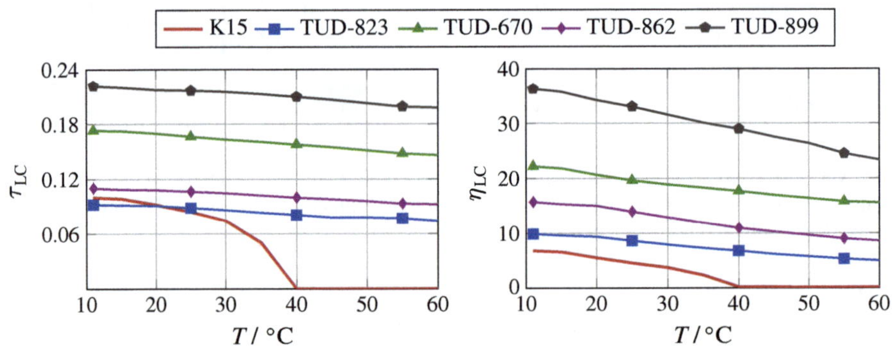

Fig. A.11 *Left* Extracted material tunabilities τ_{LC} at 20 GHz. *Right* Extracted material efficiencies η_{LC} at 20 GHz

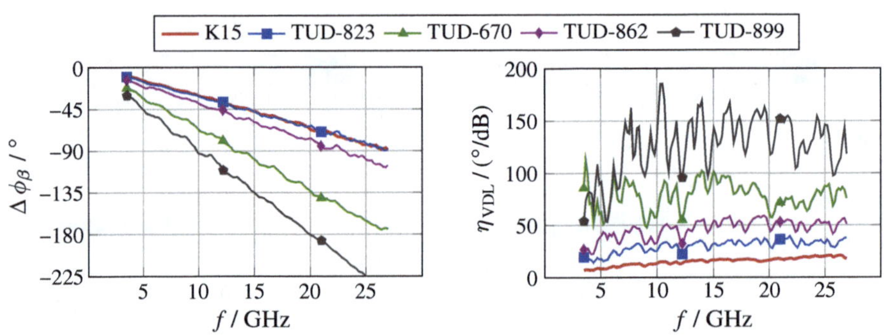

Fig. A.12 *Left* Measured differential phase shifts $\Delta\phi_\beta$ at 25 °C when the coaxial line is used as a variable delay line. *Right* The resultant delay line efficiencies η_{VDL}

A.2.2 Split-Cylinder-Cavity Method

A schematic of the split-cylinder-cavity is shown in Fig. A.13. A cylindrical cavity is split into two parts and the dielectric sample is placed between the halves [Che+04]. The advantage of this method is that there is almost no requirement for a sample preparation. Here, the approximated model [Ken96] is summarized for a sample, whose height h is

$$h \ll \frac{\lambda}{2\pi\sqrt{\varepsilon_r}}. \tag{A.7}$$

In Eq. (A.7), ε_r is the relative permittivity of the sample. The fundamental mode of the setup is TE_{011}. From the boundary conditions at the surface of the dielectric sample, one can write an eigenvalue θ equation* as [Che+04]

$$\theta \tan\theta = \frac{h}{2L}\psi \cot\psi. \tag{A.8}$$

with

$$\theta = \frac{h}{2}\sqrt{\varepsilon_r\left(\frac{2\pi}{\lambda_0}\right)^2 - \left(\frac{2\pi}{\lambda_c}\right)^2} \tag{A.9}$$

$$\psi = L\sqrt{\left(\frac{2\pi}{\lambda_0}\right)^2 - \left(\frac{2\pi}{\lambda_c}\right)^2} \tag{A.10}$$

where L is length of the half cavity, h is the thickness of the dielectric sample, λ_0 and λ_c are wavelengths at the resonant frequency and at the cutoff frequency of TE_{011} mode, respectively. equation* (A.8) can be solved for θ by using a non-linear root mean square algorithm when the resonant frequency is determined by measurement and the structural data is known. Then, the relative permittivity can be calculated as [Ken96, Che+04]

$$\varepsilon_r = \frac{1 + (\theta/\theta_c)^2}{(\lambda_c/\lambda_0)^2} \tag{A.11}$$

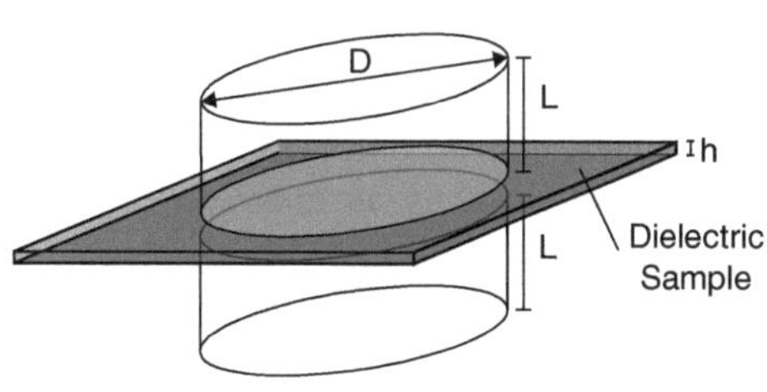

Fig. A.13 A schematic of the split-cylinder cavity [Che+04]. The dimensions of the resonator used in this work are $L = 5.5$ mm and $D = 20$ mm

with $\theta_c = j_{11}h/D$, where $j_{11} = 3.83171$ is the 1st zero of the Bessel function derivative $J_0'(x)$ and D is the diameter of the cylinder.
The loss tangent $\tan\delta$ can be calculated using [Ken96, Che+04]

$$\tan\delta = \frac{1+U}{Q_0} - SU(1+W), \tag{A.12}$$

where U and W are partition functions and S is a measure of the loss of the cavity itself and Q_0 is the loaded quality factor. The partition function U is the ratio of the electric energy in the cavity outside the dielectric sample to the electric energy in the dielectric sample. This is given by

$$U = \frac{1}{\varepsilon_r}\frac{\cos^3\theta}{(\theta + \sin\theta\cos\theta)\sin\theta}. \tag{A.13}$$

The second partition function W in Eq. (A.12) is the ratio of the end-wall loss to side wall loss is given by

$$W = j_{11}\frac{(\psi/\psi_c)^3}{\psi - \sin\psi\cos\psi} \tag{A.14}$$

with $\psi_c = j_{11}2L/D$. The final missing parameter S in Eq. (A.12) is calculated by

$$S = \frac{0.722}{\sqrt{60\pi D\sigma(\lambda_c/\lambda_0)^3}}, \tag{A.15}$$

where σ is the conductivity of the metal forming the cavity. Within the approximated model given here, it is assumed that the length of the dielectric sample is equal to the diameter of the cavity. However, in reality the length of the sample extends outside the cavity, leading to a change in the resonance frequency. Additionally, depending on the sample height and the cavity diameter a gap correction can be crucial. For further details of the gap correction one can refer [KB96].

A.3 Transmission Lines

A.3.1 Coaxial Line

In a coaxial line, an inner conductor is surrounded by a dielectric material and both are then shielded by an outer conductor [Poz11]. A schematic and its equivalent circuit are given in Fig. A.14.
The circuitry includes a serial resistance R, serial inductance L, a shunt capacitance C and a shunt conductance G. The serial resistance represents the metallic losses and the shunt conductance originates from the dielectric losses. These values can be

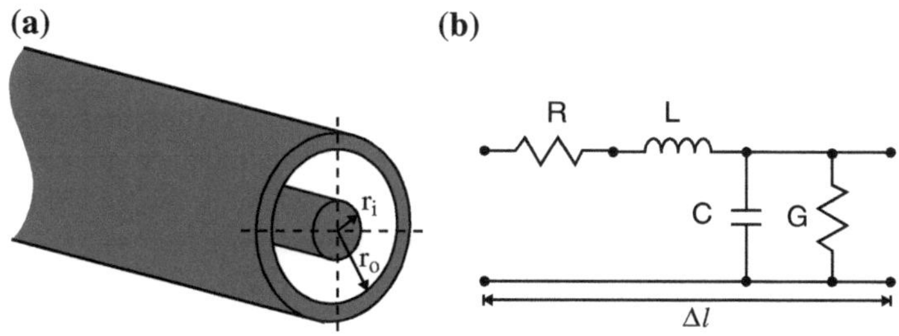

Fig. A.14 **a** Schematic and **b** equivalent circuit of a coaxial line. r_o and r_i are the inner and outer conductor radiuses, respectively

calculated quantitatively by

$$R = \frac{R_s}{2\pi}\left(\frac{1}{r_o} + \frac{1}{r_i}\right) \quad \text{with} \quad R_s = \sqrt{\frac{\omega\mu_0}{2\sigma}}, \tag{A.16}$$

$$L = \frac{\mu_0}{2\pi}\ln(r_o/r_i), \tag{A.17}$$

$$C = \frac{2\pi\varepsilon_r'\varepsilon_0}{\ln(r_o/r_i)} \tag{A.18}$$

and

$$G = \frac{2\pi\varepsilon_r\varepsilon_0\omega}{\ln(r_o/r_i)}\tan\delta \tag{A.19}$$

for per unit length when dielectric properties is $\underline{\varepsilon}_r = \varepsilon_r(1 - j\tan\delta)$. In Eqs. (A.16)–(A.19), ω is the angular frequency, σ is the conductivity of the electrodes, ε_0 and μ_0 permittivity and permeability of free space, respectively. The propagation constant $\underline{\gamma}$ and the characteristic impedance Z_{Line} of such a line can be calculated by

$$\underline{\gamma} = \sqrt{(R + j\omega L)(G + j\omega C)} \tag{A.20}$$

and

$$Z_{\text{Line}} = \sqrt{\frac{R + j\omega L}{G + j\omega C}}, \tag{A.21}$$

respectively.

A.3.2 Coplanar Waveguide

The dielectric parameters of a sheet sample are determined from the scattering parameters measurements of a coplanar waveguide (CPW), which is implemented on the dielectric sample under test. A schematic of the CPW line is shown in Fig. A.15. The effective permittivity $\varepsilon_{r,\text{eff}}$ of the CPW line is given by [Men+03]

$$\varepsilon_{r,\text{eff}} = -\left(\frac{-\ln T}{\omega \ell_{\text{physical}} \sqrt{\varepsilon_0 \mu_0}}\right), \tag{A.22}$$

where ω is the angular frequency, ε_0 and μ_0 are free space permittivity (8.85×10^{-12} F/m) and permeability ($4\pi \times 10^{-7}$ N/A^2), respectively, ℓ_{physical} is the physical length of the CPW. T is first transmission coefficient, which can be calculated by using the formulas given in A.2.1.
Once $\varepsilon_{r,\text{eff}}$ is determined, the relative permittivity of the dielectric sample ε_r can be calculated by using the following equation* [Men+03]:

$$\varepsilon_{r,\text{eff}} = 1 + q(\varepsilon_r - 1) \tag{A.23}$$

with a filling factor q [Sim01]

$$q = \frac{1}{2}\frac{K(k)K(k_1')}{K(k')K(k_1)}, \tag{A.24}$$

where K is the complete elliptical integral of the first kind and $k' = \sqrt{1-k^2}$. Here, the arguments k and k' are defined by the CPW geometry and given by

$$k = \frac{c}{b}\sqrt{\frac{b^2 - a^2}{c^2 - a^2}} \tag{A.25}$$

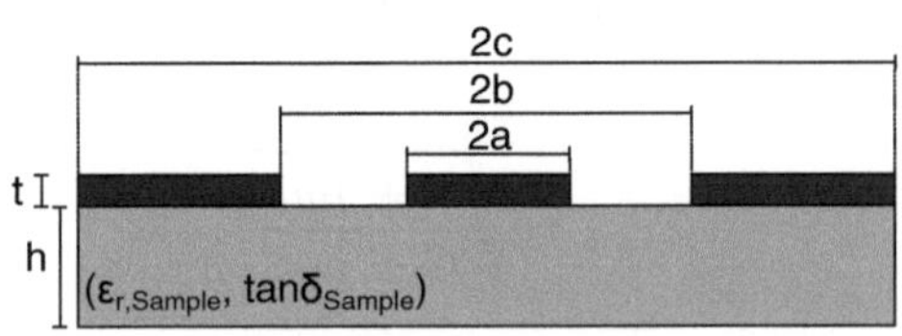

Fig. A.15 Cross-sectional view of the coplanar waveguide (CPW) [Men+03]. t is the electrode thickness and h is the height of the dielectric sample, which is defined by its relative permittivity ε_r and loss tangent $\tan\delta$

$$k_1 = \frac{\sinh x_c}{\sinh x_b}\sqrt{\frac{\sinh^2 x_b - \sinh^2 x_a}{\sinh^2 x_c - \sinh^2 x_a}}, \tag{A.26}$$

where $x_i = (\pi i/2h)$ and i is a, b or c.
The loss tangent of the sample $\tan\delta$ can be computed from the complex propagation constant $\underline{\gamma}$ of the CPW, which is given by [Men+03]

$$e^{-\underline{\gamma}\ell_{\text{physical}}} = \left(\frac{1 - S_{11}^2 + S_{21}^2}{2S_{21}} \pm K\right)^{-1} \tag{A.27}$$

with

$$K = \left(\frac{(S_{11}^2 - S_{21}^2 + 1)^2 - 4S_{11}^4}{4S_{21}^4}\right)^{1/2}. \tag{A.28}$$

Indeed, $\underline{\gamma}$ includes both losess caused by the finite conductors α_c and caused by the dielectric sample α_d (that is $\underline{\gamma} = (\alpha_c + \alpha_d) + j\beta$). The formulas for α_c are given by [HK95]

$$\alpha_c \approx \frac{R_{sm} b^2}{16 Z_0 K^2(k)(b^2 - a^2)}\left(\frac{1}{a}\ln\left(\frac{2a(b-a)}{\Delta(b+a)}\right) + \frac{1}{b}\ln\left(\frac{2b(b-a)}{\Delta(b+a)}\right)\right) \text{ (Nepers/m)} \tag{A.29}$$

with

$$R_{sm} = \omega\mu_c t\Im\left(\frac{\cot k_c t + \csc k_c t}{k_c t}\right), \tag{A.30}$$

where Z_0 is line impedance (50 Ω), k_c is the wave number, μ_c is conductor permeability with $\mu_c = \mu_0\mu_{rc}$ ($\mu_{rc,\text{copper}} \approx 1$ [Sim01]), t is conductor thickness and Δ is stopping distance. The stopping distance is determined by evaluating the power loss from an integration of the current distribution around the contour of the edge [HK95]. Once γ and α_c are computed, one can obtain α_d as [Sim01]

$$\alpha_d = \underline{\gamma} - \alpha_c = \frac{\pi\varepsilon_r}{\lambda_0\sqrt{\varepsilon_{r,\text{eff}}}} q \tan\delta \text{ (Nepers/m)}. \tag{A.31}$$

A.3.3 Multilayer Microstrip Line

The calculations of effective permittivity ε_{eff} and characteristic impedance Z_0 of a multilayer microstrip line are presented. The analytical formulas given here are obtained in [Sva92] by using a conformal mapping method. A schematic of the multilayer microstrip line is shown in Fig. A.16.
Based on Wheeler's transformation, filling factors q_1 and q_2 are given by [Sva92]

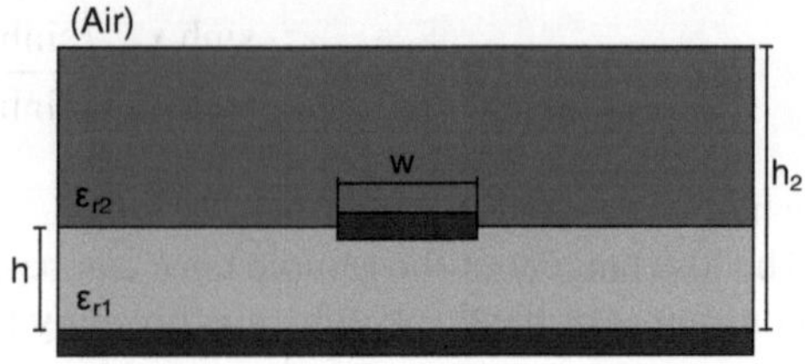

Fig. A.16 Cross-sectional view of the multilayer microstrip line [Sva92]

$$q_1 = 1 - \frac{1}{2}\frac{\ln\left(\pi\frac{w_{\text{eff}}}{h} - 1\right)}{w_{\text{eff}}/h} \tag{A.32}$$

$$q_2 = 1 - q_1 - \frac{1}{2}\frac{h - v_\varepsilon}{w_{\text{eff}}}\ln\left(\pi\frac{w_{\text{eff}}}{h}\frac{\cos\left(\frac{v_\varepsilon}{2}\frac{\pi}{h}\right)}{\pi\left(\frac{h_2}{h} - 0.5\right) + \frac{v_\varepsilon}{2}\frac{\pi}{h}} + \sin\left(\frac{v_\varepsilon}{2}\frac{\pi}{h}\right)\right) \tag{A.33}$$

for $w/h \geq 1$ with

$$w_{\text{eff}} = w + \frac{2h}{\pi}\ln\left(17.08\left(\frac{w}{2h} + 0.92\right)\right) \tag{A.34}$$

$$v_\varepsilon = 2\frac{h}{\pi}\text{arctg}\left(\frac{\pi}{\frac{\pi}{2}\frac{w_{\text{eff}}}{h} - 2}\left(\frac{h_2}{h} - 1\right)\right) \tag{A.35}$$

Similarly, the filling factors are

$$q_1 = \frac{1}{2} + \frac{0.9}{\pi\ln\frac{h}{w}} \tag{A.36}$$

$$q_2 = \frac{1}{2} - \frac{0.9 + \frac{\pi}{4}\ln(K)\arccos\left(\left(1 - \frac{h}{h_2}\left(1 - \frac{w}{8h}\right)\right)\sqrt{K}\right)}{\pi\ln\frac{8h}{w}} \tag{A.37}$$

for $w/h \leq 1$ with $K = (h_2/h + 1)/(h_2/h + w/4h - 1)$. Then, the effective permittivity of the multilayer microstrip line is calculated by

$$\varepsilon_{r,\text{eff}} = \varepsilon_{r1}q_1 + \varepsilon_{r2}\frac{(1 - q_1)^2}{\varepsilon_{r2}(1 - q_1 - q_2) + q_2}. \tag{A.38}$$

The corresponding characteristic impedance Z_0 is given by

$$Z_0 = \frac{120\pi}{\sqrt{\varepsilon_{r,\text{eff}}}} \frac{h}{w_{\text{eff}}} \tag{A.39}$$

for $w/h \geq 1$ and

$$Z_0 = \frac{60}{\sqrt{\varepsilon_{r,\text{eff}}}} \ln \frac{8h}{w} \tag{A.40}$$

for $w/h \leq 1$.
Attenuation caused by the dielectric loss is determined by [Poz11]

$$\alpha_d = \frac{\pi \varepsilon_{r1}(\varepsilon_{r,\text{eff}} - 1) \tan \delta}{\lambda_0 \sqrt{\varepsilon_{r,\text{eff}}}(\varepsilon_{r1} - 1)} \text{ (Nepers/m)} \tag{A.41}$$

for a conventional microstrip line ($\varepsilon_{r2} = 1, h_2 - h \to \infty$). In Eq. (A.41), λ_0 is the wavelength and $\tan \delta$ is the loss tangent of the dielectric. Attenuation caused by the conductor loss is estimated by

$$\alpha_c \approx \frac{R_s}{Z_0 w} \text{ (Nepers/m)}, \tag{A.42}$$

where R_s is the surface resistivity that is $R_s = \sqrt{\omega\mu_0/2\sigma}$. ω is the angular frequency, μ_0 is the permeability of free space ($4\pi \times 10^{-7}\,\text{N/A}^2$) and σ is the conductivity of the conductor.

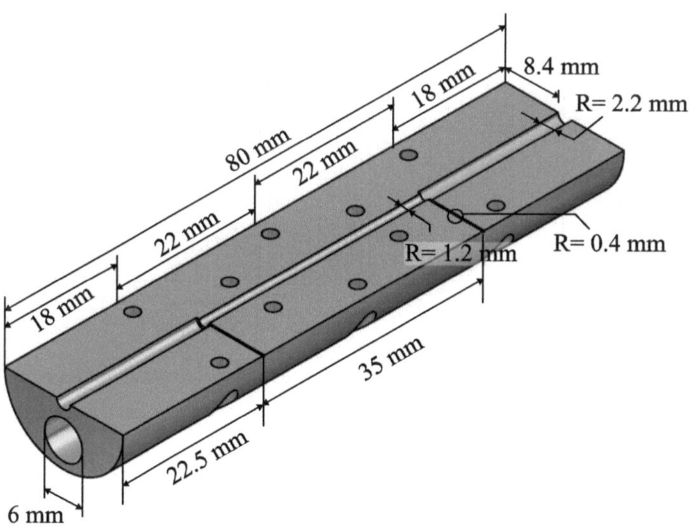

Fig. A.17 Dimensional details of the split blocks used in the coaxial line method

A.4 Layouts and Dimensions

A.4.1 Coaxial Line Method

See Fig. A.17.

A.4.2 Inverted Microstrip Delay Line

See Fig. A.18.

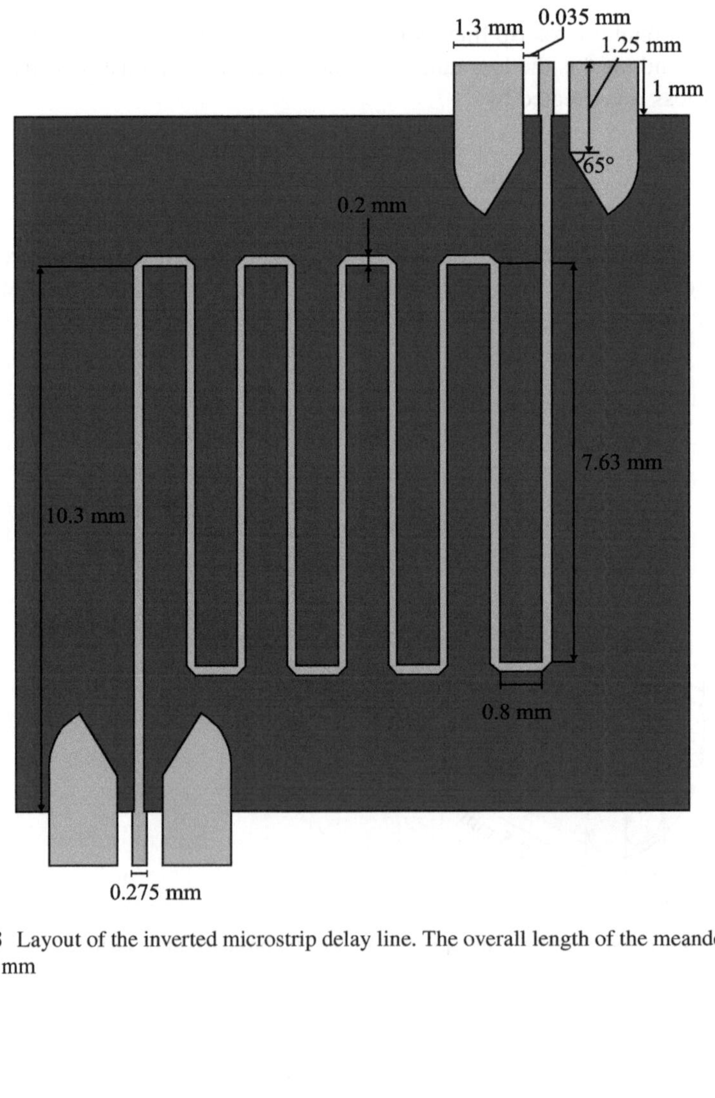

Fig. A.18 Layout of the inverted microstrip delay line. The overall length of the meandered delay line is 75 mm

A.4.3 Loaded Coplanar Waveguide Delay Line

See Fig. A.19.

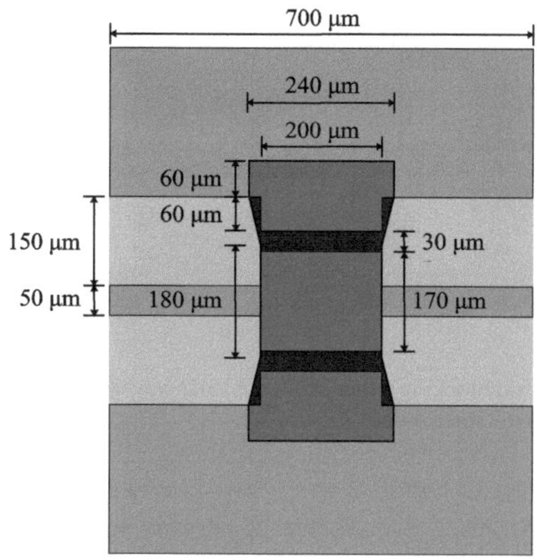

Fig. A.19 Layout of an unit element of the loaded CPW delay line. The unit elements consists of not tunable CPW lines and a shunt LC varactor in between

A.4.4 2D Electronic Beam Steering Phased Array

Single Radiating Element

See Fig. A.20.

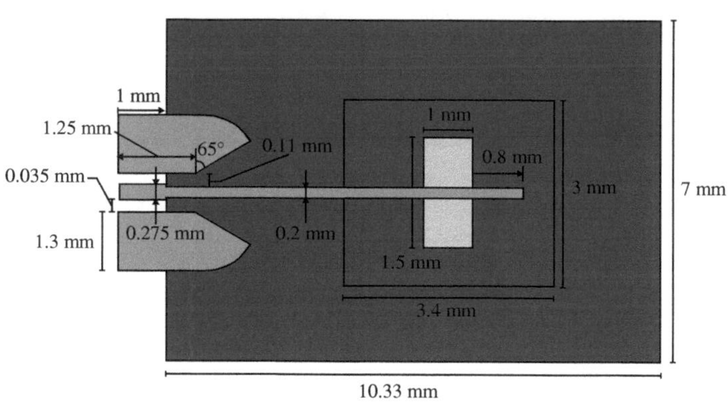

Fig. A.20 Layout of the single radiating element. The structure includes an aperture coupled microstrip patch antenna

DC-Blocking Structure

See Fig. A.21.

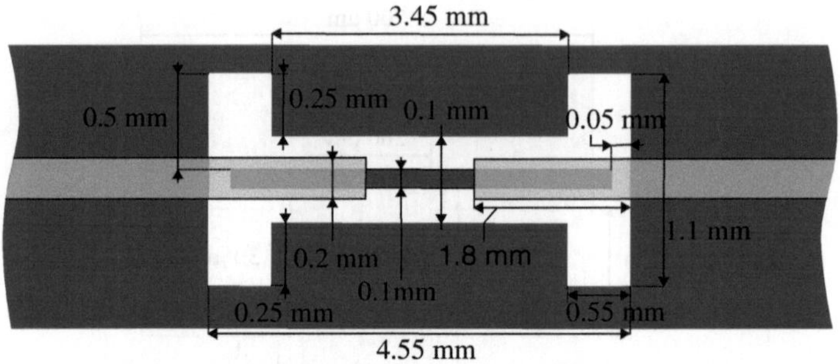

Fig. A.21 Layout of the dc-blocking structure. A coplanar waveguide is implemented in the ground electrode of the inverted microstrip line

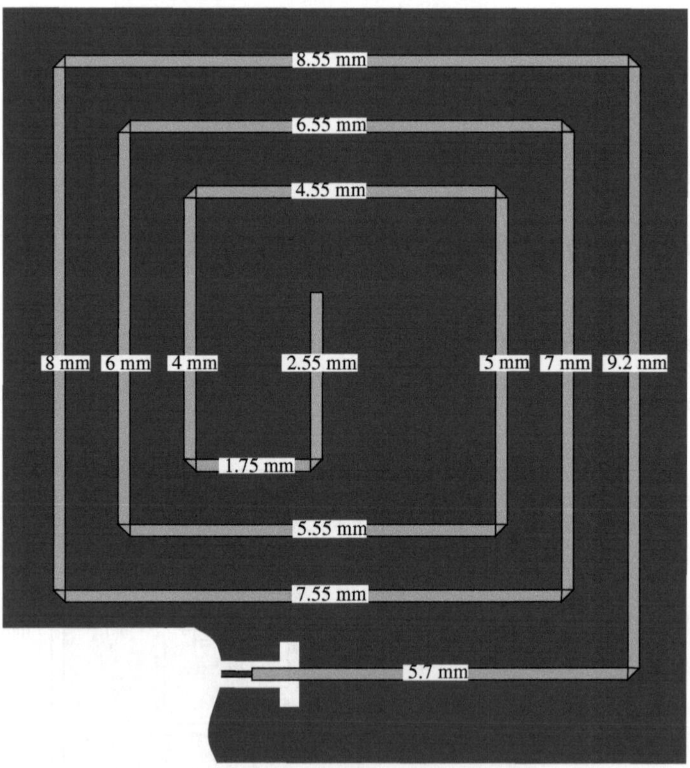

Fig. A.22 Layout of the inverted microstrip delay line in the spiral geometry. The overall length of the line is 80 mm

Spiral Delay Line

See Fig. A.22.

A.4.5 Tunable Coupled Line based Polarization Agile Antenna

See Fig. A.23.

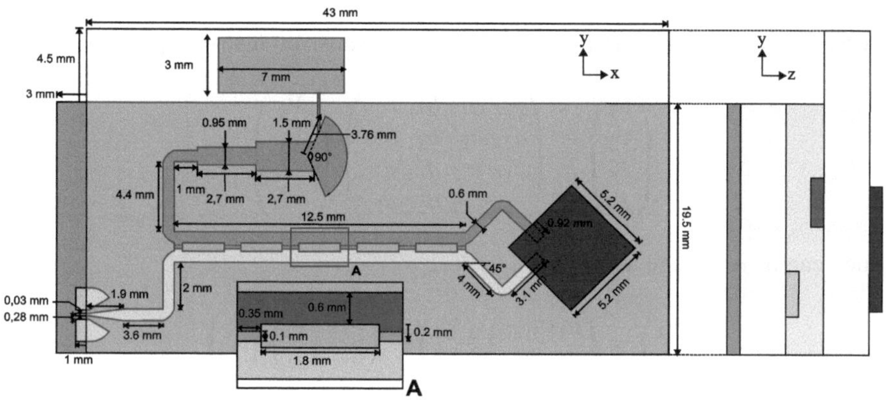

Fig. A.23 Layout of the tunable coupled line based polarization agile antenna. The coupled line includes five unit elements, where the one is highlighted. It is connected to a microstrip patch antenna through a proximity coupling

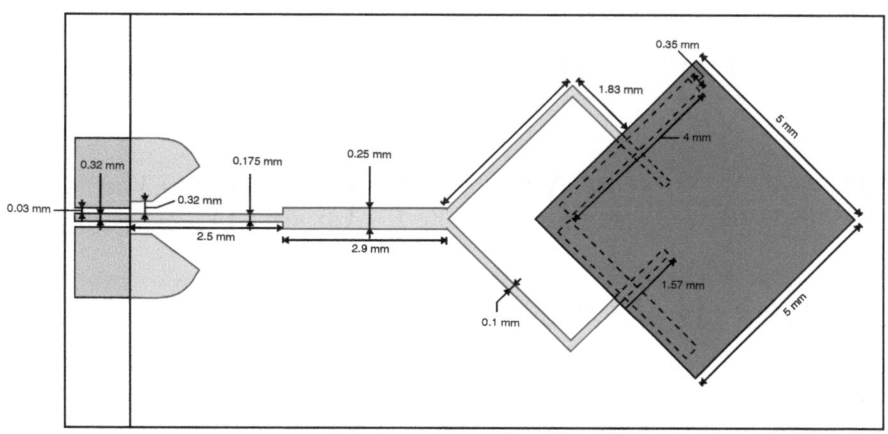

Fig. A.24 Layout of the single radiating element, which is fabricated using the glass and printed circuit board

A.4.6 Variable Delay Line based Polarization Agile Antenna

See Fig. A.24.

A.5 Convertion of 4-Port Transmission (ABCD) Matrix Into 4-Port Scattering Matrix

This part presents the derivations of the formulas converting a 4-port transmission matrix into a scattering parameters. A 4-port transmission matrix is given as

$$\begin{pmatrix} V_1 \\ V_3 \\ I_1 \\ I_3 \end{pmatrix} = \begin{pmatrix} a_{11} & a_{12} & b_{11} & b_{12} \\ a_{21} & a_{22} & b_{21} & b_{22} \\ c_{11} & c_{12} & d_{11} & d_{12} \\ c_{21} & c_{22} & d_{21} & d_{22} \end{pmatrix} \begin{pmatrix} V_2 \\ V_4 \\ -I_2 \\ -I_4 \end{pmatrix}, \tag{A.43}$$

which can be rearrange as

$$\begin{pmatrix} V_1 \\ V_3 \end{pmatrix} = \begin{pmatrix} a_{11} & a_{12} \\ a_{21} & a_{22} \end{pmatrix} \begin{pmatrix} V_2 \\ V_4 \end{pmatrix} - \begin{pmatrix} b_{11} & b_{12} \\ b_{21} & b_{22} \end{pmatrix} \begin{pmatrix} I_2 \\ I_4 \end{pmatrix}, \tag{A.44}$$

$$\begin{pmatrix} I_1 \\ I_3 \end{pmatrix} = \begin{pmatrix} c_{11} & c_{12} \\ c_{21} & c_{22} \end{pmatrix} \begin{pmatrix} V_2 \\ V_4 \end{pmatrix} - \begin{pmatrix} d_{11} & d_{12} \\ d_{21} & d_{22} \end{pmatrix} \begin{pmatrix} I_2 \\ I_4 \end{pmatrix}. \tag{A.45}$$

The potential at a port is actually superposition of an incident voltage (V^+) and a leaving voltage (V^-). Hence, Eqs. (A.44) and (A.45) are written as

$$\begin{pmatrix} V_1^+ + V_1^- \\ V_3^+ + V_3^- \end{pmatrix} = \begin{pmatrix} a_{11} & a_{12} \\ a_{21} & a_{22} \end{pmatrix} \begin{pmatrix} V_2^+ + V_2^- \\ V_4^+ + V_4^- \end{pmatrix} - Y_0 \begin{pmatrix} b_{11} & b_{12} \\ b_{21} & b_{22} \end{pmatrix} \begin{pmatrix} V_2^+ - V_2^- \\ V_4^+ - V_4^- \end{pmatrix} \tag{A.46}$$

$$\begin{pmatrix} V_1^+ - V_1^- \\ V_3^+ - V_3^- \end{pmatrix} = Z_0 \begin{pmatrix} c_{11} & c_{12} \\ c_{21} & c_{22} \end{pmatrix} \begin{pmatrix} V_2^+ + V_2^- \\ V_4^+ + V_4^- \end{pmatrix} - \begin{pmatrix} d_{11} & d_{12} \\ d_{21} & d_{22} \end{pmatrix} \begin{pmatrix} V_2^+ - V_2^- \\ V_4^+ - V_4^- \end{pmatrix} \tag{A.47}$$

$$\begin{pmatrix} V_1^+ \\ V_3^+ \end{pmatrix} + \begin{pmatrix} V_1^- \\ V_3^- \end{pmatrix} = (A - Y_0 B) \begin{pmatrix} V_2^+ \\ V_4^+ \end{pmatrix} + (A + Y_0 B) \begin{pmatrix} V_2^- \\ V_4^- \end{pmatrix}, \tag{A.48}$$

$$\begin{pmatrix} V_1^+ \\ V_3^+ \end{pmatrix} - \begin{pmatrix} V_1^- \\ V_3^- \end{pmatrix} = (Z_0 C - D) \begin{pmatrix} V_2^+ \\ V_4^+ \end{pmatrix} + (Z_0 C + D) \begin{pmatrix} V_2^- \\ V_4^- \end{pmatrix}, \tag{A.49}$$

where

$$A = \begin{pmatrix} a_{11}a_{12} \\ a_{21}a_{22} \end{pmatrix}, \tag{A.50}$$

$$B = \begin{pmatrix} b_{11}b_{12} \\ b_{21}b_{22} \end{pmatrix}, \tag{A.51}$$

$$C = \begin{pmatrix} c_{11}c_{12} \\ c_{21}c_{22} \end{pmatrix}, \tag{A.52}$$

$$D = \begin{pmatrix} d_{11}d_{12} \\ d_{21}d_{22} \end{pmatrix}. \tag{A.53}$$

One can obtain the following equation*s by adding and the subtracting Eqs. (A.48) and (A.49):

$$2\begin{pmatrix} V_1^+ \\ V_3^+ \end{pmatrix} = (A - Y_0B + Z_0C - D)\begin{pmatrix} V_2^+ \\ V_4^+ \end{pmatrix} + (A + Y_0 + Z_0C + D)\begin{pmatrix} V_2^- \\ V_4^- \end{pmatrix} \tag{A.54}$$

$$2\begin{pmatrix} V_1^- \\ V_3^- \end{pmatrix} = (A - Y_0B - Z_0C + D)\begin{pmatrix} V_2^+ \\ V_4^+ \end{pmatrix} + (A + Y_0 - Z_0C - D)\begin{pmatrix} V_2^- \\ V_4^- \end{pmatrix} \tag{A.55}$$

equation*s (A.54) and (A.55) can be further simplified as

$$2\begin{pmatrix} V_1^+ \\ V_3^+ \end{pmatrix} = \begin{pmatrix} e_{11} & e_{12} \\ e_{22} & e_{22} \end{pmatrix}\begin{pmatrix} V_2^+ \\ V_4^+ \end{pmatrix} + \begin{pmatrix} f_{11} & f_{12} \\ f_{21} & f_{22} \end{pmatrix}\begin{pmatrix} V_2^- \\ V_4^- \end{pmatrix}, \tag{A.56}$$

$$2\begin{pmatrix} V_1^- \\ V_3^- \end{pmatrix} = \begin{pmatrix} g_{11} & g_{12} \\ g_{22} & g_{22} \end{pmatrix}\begin{pmatrix} V_2^+ \\ V_4^+ \end{pmatrix} + \begin{pmatrix} h_{11} & h_{12} \\ h_{21} & h_{22} \end{pmatrix}\begin{pmatrix} V_2^- \\ V_4^- \end{pmatrix}, \tag{A.57}$$

where

$$A - Y_0B + Z_0C - D = \begin{pmatrix} e_{11}e_{12} \\ e_{21}e_{22} \end{pmatrix}, \tag{A.58}$$

$$A + Y_0 + Z_0C + D = \begin{pmatrix} b_{11}b_{12} \\ b_{21}b_{22} \end{pmatrix}, \tag{A.59}$$

$$A - Y_0 B - Z_0 C + D = \begin{pmatrix} c_{11} c_{12} \\ c_{21} c_{22} \end{pmatrix}, \tag{A.60}$$

$$A + Y_0 - Z_0 C - D = \begin{pmatrix} d_{11} d_{12} \\ d_{21} d_{22} \end{pmatrix}. \tag{A.61}$$

The incident and leaving voltages are rearranged as

$$f_{11} V_2^- + f_{12} V_4^- = 2V_1^+ - e_{11} V_2^+ - e_{12} V_4^+ \tag{A.62}$$

$$f_{21} V_2^- + f_{22} V_4^- = 2V_1^+ - e_{21} V_2^+ - e_{22} V_4^+ \tag{A.63}$$

$$2V_1^- = g_{11} V_2^+ + g_{12} V_4^+ + h_{11} V_2^- + h_{12} V_4^- \tag{A.64}$$

$$2V_3^- = g_{21} V_2^+ + g_{22} V_4^+ + h_{21} V_2^- + h_{22} V_4^- \tag{A.65}$$

to obtain

$$V_2^- = \frac{2f_{22} V_1^+ - 2f_{12} V_3^+ - (e_{11} f_{22} - e_{21} f_{12}) V_2^+ - (e_{12} f_{22} - e_{22} f_{12}) V_4^+}{f_{11} f_{22} - f_{12} f_{21}}, \tag{A.66}$$

$$V_4^- = \frac{2f_{21} V_1^+ - 2f_{11} V_3^+ - (e_{11} f_{21} - e_{21} f_{11}) V_2^+ - (e_{12} f_{21} - e_{22} f_{11}) V_4^+}{f_{11} f_{22} - f_{12} f_{21}}, \tag{A.67}$$

$$\begin{aligned} V_1^- &= \frac{h_{11} f_{22} - h_{12} f_{21}}{f_{11} f_{22} - f_{12} f_{21}} V_1^+ \\ &\quad + 0.5\left(g_{11} - \frac{h_{11}(e_{11} f_{22} - e_{21} f_{12}) - h_{12}(e_{11} f_{21} - e_{21} f_{11})}{f_{11} f_{22} - f_{12} f_{21}} \right) V_2^+ \\ &\quad + \frac{h_{12} f_{11} - h_{11} f_{12}}{f_{11} f_{22} - f_{12} f_{21}} V_3^+ \\ &\quad + 0.5\left(g_{12} - \frac{h_{11}(e_{12} f_{22} - e_{22} f_{12}) - h_{12}(e_{12} f_{21} - e_{22} f_{11})}{f_{11} f_{22} - f_{12} f_{21}} \right) V_4^+, \end{aligned} \tag{A.68}$$

$$V_3^- = \frac{f_{22}h_{21} - f_{21}h_{22}}{f_{11}f_{22} - f_{12}f_{21}} V_1^+ \\ + \left(g21 - \frac{h_{21}(e_{11}f_{22} - e_{21}f_{12}) - h_{22}(e_{11}f_{21} - e_{21}f_{11})}{f_{11}f_{22} - f_{12}f_{21}} \right) V_2^+ \\ + \frac{h_{22}f_{11} - h_{21}f_{12}}{f_{11}f_{22} - f_{12}f_{21}} V_3^+ \\ + \left(g_{22} - \frac{h_{21}(e_{12}f_{22} - e_{22}f_{12}) - h_{22}(e_{12}f_{21} - e_{22}f_{11})}{f_{11}f_{22} - f_{12}f_{21}} \right) V_4^+. \tag{A.69}$$

Finally, the indices of the resultant scattering parameters can be calculated by using the definition of the scattering parameters [Poz11] which is $S_{ij} = V_i^- / V_j^+$ when all other incident voltages are equal to 0 V.

$$S_{11} = \frac{h_{11}f_{22} - h_{12}f_{21}}{f_{11}f_{22} - f_{12}f_{21}}, \tag{A.70}$$

$$S_{12} = 0.5 \left(g_{11} - \frac{h_1 1(e_{11}f_{22} - e_{21}f_{12}) - h_{12}(e_{11}f_{21} - e_{21}f_{11})}{f_{11}f_{22} - f_{12}f_{21}} \right), \tag{A.71}$$

$$S_{13} = \frac{h_{12}f_{11} - h_{11}f_{12}}{f_{11}f_{22} - f_{12}f_{21}}, \tag{A.72}$$

$$S_{14} = 0.5 \left(g_{12} - \frac{h_{11}(e_{12}f_{22} - e_{22}f_{12}) - h_{12}(e_{12}f_{21} - e_{22}f_{11})}{f_{11}f_{22} - f_{12}f_{21}} \right), \tag{A.73}$$

$$S_{21} = \frac{2f_{22}}{f_{11}f_{22} - f_{12}f_{21}}, \tag{A.74}$$

$$S_{22} = \frac{e_{21}f_{12} - e_{11}f_{22}}{f_{11}f_{22} - f_{12}f_{21}}, \tag{A.75}$$

$$S_{23} = \frac{-2f_{12}}{f_{11}f_{22} - f_{12}f_{21}}, \tag{A.76}$$

$$S_{24} = \frac{e_{22}f_{12} - e_{12}f_{22}}{f_{11}f_{22} - f_{12}f_{21}}, \tag{A.77}$$

$$S_{31} = \frac{h_{21} f_{22} - h_{22} f_{21}}{f_{11} f_{22} - f_{12} f_{21}}, \tag{A.78}$$

$$S_{32} = 0.5 \left(g_{21} - \frac{h_{21}(e_{11} f_{22} - e_{21} f_{12}) - h_{22}(e_{11} f_{21} - e_{21} f_{11})}{f_{11} f_{22} - f_{12} f_{21}} \right), \tag{A.79}$$

$$S_{33} = \frac{h_{22} f_{11} - h_{21} f_{12}}{f_{11} f_{22} - f_{12} f_{21}}, \tag{A.80}$$

$$S_{34} = 0.5 \left(g_{22} - \frac{h_{21}(e_{12} f_{22} - e_{22} f_{12}) - h_{22}(e_{12} f_{21} - e_{22} f_{11})}{f_{11} f_{22} - f_{12} f_{21}} \right), \tag{A.81}$$

$$S_{41} = \frac{-2 f_{21}}{f_{11} f_{22} - f_{12} f_{21}}, \tag{A.82}$$

$$S_{42} = \frac{e_{11} f_{21} - e_{21} f_{11}}{f_{11} f_{22} - f_{12} f_{21}}, \tag{A.83}$$

$$S_{43} = \frac{2 f_{11}}{f_{11} f_{22} - f_{12} f_{21}}, \tag{A.84}$$

$$S_{44} = \frac{e_{12} f_{21} - e_{22} f_{11}}{f_{11} f_{22} - f_{12} f_{21}}. \tag{A.85}$$

References

[BJ+93] J. Baker-Jarvis, M.D. Janezic, J.H. Grosvenor, R.G. Geyer, *Transmission/reflection and short-circuit line methods for measuring permittivity and permeability*. Technical report, National Institute of Standards and Technology, 1355-R, 1993

[Che+04] L.F. Chen, C.K. Ong, C.P. Neo, V.V. Varadan, V.K. Varadan, *Microwave Electronics: Measurement and Materials Characterization* (Wiley, New York, 2004)

[Goe09] F. Gölden, Liquid Crystal Based Microwave Components with Fast Response Times: Material, Technology, Power Handling Capability. PhD thesis, Technische Universität Darmstadt, 2009

[HK95] C.L. Holloway, E.F. Kuester, A quasi-closed form expression for the conductor loss of CPW lines, with an investigation of edge shape effects. IEEE Trans. Microw. Theory Tech. **43**, 2695–2701 (1995)

[Ken96] G. Kent, Nondestructive permittivity measurement of substrates. IEEE Trans. Instr. Meas. **45**, 102–106 (1996)

[KB96] G. Kent, S.M. Bell, The gap correction for the resonant-mode dielectrometer. IEEE Trans. Instr. Meas. **45**, 98–101 (1996)

[Mic] MicroChemicals, *Product Overview*, http://www.microchemicals.eu/products.html

[Men+03] P.M. Mendes, A. Polyakov, M. Bartek, J.N. Burghartz, J.H. Correia, Extraction of glass-wafers electrical properties based on s-parameters measurements of coplanar waveguides, in *Conftele. Conference on Telecommunications*. Instituto de Telecomunicaões, 2003, http://hdl.handle.net/1822/1653

[NR70] A.M. Nicolson, G.F. Ross, Measurement of the intrinsic properties of materials by time-domain techniques. IEEE Trans. Instrum. Meas. **19**(4), 377–382 (1970)

[Poz11] D.M. Pozar, *Microwave Engineering*, 4th edn. (Wiley, New York, 2011)

[Sim01] R.N. Simons, *Coplanar Waveguide Circuits, Components, and Systems* (Wiley, New York, 2001)

[Sva92] J. Svacina, Analysis of multilayer microstrip lines by a conformal mapping method. IEEE Trans. Microw. Theory Tech. **40**, 769–772 (1992)

[Wei74] W.B. Weir, Automatic measurement of complex dielectric constant and permeability at microwave frequencies. Proc. IEEE **62**(1), 33–36 (1974)

About the Author

Dr.-Ing. Onur Hamza Karabey was born in Istanbul, Turkey, in 1983. He received his B.Sc. degree from the Middle East Technical University, Ankara, Turkey, in 2006, M.Sc. and Ph.D. degrees from the Technische Universität Darmstadt, Germany, in 2008, and 2013, respectively, all in electrical and electronics engineering. His current research is focused on broadband material characterization for microwaves and design of tunable passive components for reconfigurable antennas in liquid crystal technology, which is used as tunable dielectric. Dr. Karabey won the "Ideenwettbewerb 2011" of Technische Universität Darmstadt in 2012 because of realizing a two-dimensional electronic beam steering antenna using the liquid crystal display technology for the first time. He also involved in the scientific work and in a technical paper as a co-author, which won the Best Paper Award of the European Microwave Integrated Circuit Conference in 2011.

O. H. Karabey, *Electronic Beam Steering and Polarization Agile Planar Antennas in Liquid Crystal Technology*, Springer Theses, DOI: 10.1007/978-3-319-01424-1,

MIX
Papier aus verantwortungsvollen Quellen
Paper from responsible sources
FSC® C105338

If you have any concerns about our products,
you can contact us on
ProductSafety@springernature.com

In case Publisher is established outside the EU,
the EU authorized representative is:
Springer Nature Customer Service Center GmbH
Europaplatz 3, 69115 Heidelberg, Germany

Printed by Libri Plureos GmbH
in Hamburg, Germany